T0225110

◆ Revised Edition ◆

UNDERSTANDING BASIC CHEMISTRY THROUGH PROBLEM SOLVING

THE LEARNER'S APPROACH

Revised Edition

UNDERSTANDING BASIC CHEMISTRY THROUGH PROBLEM SOLVING

THE LEARNER'S APPROACH

KIM SENG CHAN

JEANNE TAN

WS Education

NEW JERSEY · LONDON · SINGAPORE · BEIJING · SHANGHAI · HONG KONG · TAIPEI · CHENNAI · TOKYO

Published by

WS Education, an imprint of
World Scientific Publishing Co. Pte. Ltd.
5 Toh Tuck Link, Singapore 596224
USA office: 27 Warren Street, Suite 401-402, Hackensack, NJ 07601
UK office: 57 Shelton Street, Covent Garden, London WC2H 9HE

Library of Congress Cataloging-in-Publication Data
Names: Chan, Kim Seng. | Tan, Jeanne.
Title: Understanding basic chemistry through problem solving /
    by Kim Seng Chan (Victoria Junior College, Singapore), Jeanne Tan.
Description: Revised edition. | Hackensack, NJ : World Scientific, 2017. |
    "Written for students taking either the University of Cambridge O-level
    examinations or the GCSE examinations"--Preface. | Includes index.
Identifiers: LCCN 2016059284 | ISBN 9789813209770 (softcover : alk. paper)
Subjects: LCSH: Chemistry--Great Britain--Textbooks. | Chemistry--Textbooks. |
    Chemistry--Great Britain--Examinations, questions, etc.--Study guides. |
    Chemistry--Examinations, questions, etc.--Study guides.
Classification: LCC QD31.3 .C374 2017 | DDC 540--dc23
LC record available at https://lccn.loc.gov/2016059284

British Library Cataloguing-in-Publication Data
A catalogue record for this book is available from the British Library.

Typeset by Stallion Press
Email: enquiries@stallionpress.com

Printed in Singapore

PREFACE

When a major examination approaches, students would start going around in search for guidebooks that can help them to consolidate the important concepts that are necessary to meet the requirements of these assessments in the shortest amount of time. Unfortunately, most guidebooks are of the expository and non-refutational type, presenting facts rather than explaining them. In addition, the links between concepts are often not made explicit and presupposes that learners would be able to make the necessary integration with the multitude of concepts that they have come across in their few years of chemical education, forgetting that some of them may lack the prior knowledge and metacognitive skills to do it meaningfully. Hence, learners would at most be able to reproduce the information that is structured and organized by the guidebook writer, but not able to construct a meaningful conceptual mental model for oneself. As a result, they would not be able to fluidly apply what they should know across different contextual questions that appear when sitting for that major examination.

This revised edition is a continuation of our previous few books — *Understanding Advanced Physical Inorganic Chemistry*, *Understanding Advanced Organic and Analytical Chemistry*, *Understanding Advanced Chemistry Through Problem Solving*, and *Understanding Basic Chemistry*, retaining the main refutational characteristics of the previous books by strategically planting think-aloud questions to promote conceptual understanding, knowledge construction, reinforcement of important concepts, and discourse opportunities. It is hoped that these essential questions would make learners be more aware of the possible conflict between their prior knowledge, which may be counterintuitive or misleading, with those presented in the text, and hence in the process, make the necessary conceptual changes. In essence, we are trying to effect metaconceptual awareness — awareness of the theoretical nature of one's thinking — while learners are trying to master the essential chemistry concepts and be more familiar with their applications in problem solving. We hope that by pointing out the differences between possible misconceptions and the actual chemistry content, we can promote such metaconceptual awareness and thus assist the learner to construct a meaningful conceptual model of understanding to meet the necessary assessment criteria. We want our learners to not only know what they know, but at the same time, have a sense of how they know what they know and how their new learnings are interrelated within the discipline. This would enable them to better appreciate and easily apply what they have learned in any novel question that they come across in major examinations.

Lastly, the content of this book would be both informative and challenging to the practices of teachers. This book would certainly illuminate the instruction of all chemistry teachers who strongly believe in teaching chemistry in a meaningful and integrative approach, from the learners' perspective. The integrated questions that are used as problem-solving tools would definitely prove useful to students in helping them revise fundamental concepts learned from previous chapters, and also grasp the importance and relevancy in the application to their current learning. Collectively, this book offers a vision of understanding and applying chemistry meaningfully and fundamentally from the learners' approach and to fellow chemistry teachers, we hope that it would help you develop a greater insight into what makes you tick, explain, enthuse, and develop in the course of your teaching.

ACKNOWLEDGEMENTS

We would like to express our sincere thanks to the staff at World Scientific Publishing Co. Pte. Ltd. for the care and attention which they have given to this book, particularly our editors Lim Sook Cheng and Sandhya Devi, our editorial assistant Chow Meng Wai and Stallion Press.

Special appreciation goes to Ms Ek Soo Ben (Principal of Victoria Junior College), Mr Cheong Tien Beng, Mrs Foo Chui Hoon, Mrs Toh Chin Ling and Mrs Ting Hsiao Shan for their unwavering support to Kim Seng Chan.

Special thanks go to all our students who have made our teaching of chemistry fruitful and interesting. We have learnt a lot from them just as they have learnt some good chemistry from us.

Finally, we thank our families for their wholehearted support and understanding throughout the period of writing this book. We would like to share with all the passionate learners of chemistry two important quotes from the *Analects of Confucius*:

學而時習之，不亦悅乎? **(Isn't it a pleasure to learn and practice what is learned time and again?)**

學而不思則罔，思而不學則殆 **(Learning without thinking leads to confusion, thinking without learning results in wasted effort)**

Kim Seng Chan
BSc (Hons), PhD, PDGE (Sec)
MEd, MA (Ed Mgt), MEd (G Ed), MEd (Dev Psy)

Jeanne Tan
BSc (Hons), PDGE (Sec), MEd (LST)

CONTENTS

CHAPTER 1

THE PARTICULATE NATURE OF MATTER

1. List the essential differences between a chemical and a physical change. Indicate what type of changes take place in the following process and explain clearly in each case:

Explanation:

During a physical change, the physical property of matter, such as temperature, pressure, density, mass, volume, color, boiling point, melting point, energy content, etc., has changed BUT the chemical property of the matter stays intact. Such physical change is *reversible* by changing the physical conditions, such as temperature and pressure, back to their original states.

During a chemical change, the chemical property of matter has changed. This change results in the formation of new substances which have different chemical compositions from the starting substance. Such chemical change is usually *irreversible* by any simple change of the physical conditions.

Q What is a chemical property?

A: The chemical property of a matter is actually its *unique* chemical potential in reacting with other matter. For example, when sodium reacts with chlorine to form sodium chloride, sodium atom loses an electron while chlorine atom gains an electron. The potential of the sodium atom to lose an electron in the *presence* of the chlorine atom is the chemical property for the sodium, while the potential to gain an electron for the chlorine atom in the *presence* of the sodium atom is the chemical property for chlorine.

Q So, does that mean that chlorine atom would forever have the same potential to gain electrons irrespective of the type of substances that it reacts with?

A: No! When a chlorine atom reacts with a hydrogen atom, the chlorine atom does not gain an electron. In fact, the chlorine atom shares electrons with the hydrogen atom. In essence, the chemical property of a matter can vary, depending on the chemical property of the other matter that it is reacting with. You can refer to *Understanding Basic Chemistry* by K.S. Chan and J. Tan for more details.

Do you know?

— Matter is made up of very small particles, such as an *atom, ion,* or *molecule,* being attracted to one another by electrostatic forces of attraction. The different strengths of the attractive forces between these small particles result in the different physical *states of matter.*

— There are three states of matter: solid, liquid, and gas. The strength of the attractive force between the particles would give rise to the physical state of the matter.

— If the strength of attraction is very strong, we have solid matter, which results in its *fixed shape* and *fixed volume.* A weaker strength of attraction gives us liquid, which has a *fixed volume* but *no fixed shape.* The strength of attraction is the weakest in a gas, causing it to have *neither a fixed shape nor a fixed volume.*

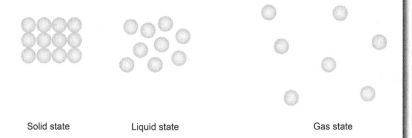

Solid state   Liquid state   Gas state

— The force that causes the small particles to be attracted together is known as *electrostatic force.* It is just a type of "electrical force" between particles that possess opposite "electrical charges." This electrostatic force is the chemical bond that holds the particles together.

(Continued)

> *(Continued)*
>
> — Thus, if the chemical bond is of *different strengths*, it will result in *different types of arrangement* of particles in the physical state, which would therefore decide whether the matter has fixed shape or fixed volume. And because of the different types of arrangement, the particles have different types of motion in the physical state and also different levels of compressibility. Furthermore, the reason for different amounts of energy being involved in the chemical reaction is due to the reaction of different types of matter possessing different chemical bond strengths.
> — In the solid state, the particles can only vibrate about a fixed position. But in the liquid and gaseous states, the particles have translational motion; it is free to move randomly in all directions.

Why is the compressibility of the solid and liquid so much lower than that for the gas?

A: Since the gas particles are far apart, when we compress a gas by exerting a pressure, the distance of separation between the gas particles can be decreased. If the pressure is high enough, the gas would be converted to the liquid state in which the particles are closer to each other. This is what happens during the *liquefaction of gas* by the application of pressure. Further compression would then be difficult to make the particles in the liquid state to come even closer together as there would be too much repulsive forces acting between the particles. We could expect more difficulty in compressing a solid than a liquid.

So, it is actually the weak attractive forces between the particles in the gaseous state that help to "pull" the particles even closer together during compression?

A: You are right! A lot of students think that it is the applied pressure ALONE that helps to "push" the particles together; this is INCORRECT. Without the attractive forces acting between the particles in the gaseous state, the particles would not be "held on" together in the liquid state. This also explains why a gas with extremely weak attractive forces is difficult to be liquefied through the using of high pressure alone.

(a) Addition of potassium to water.

Explanation:

Potassium reacts with water to give potassium hydroxide and hydrogen as follows:

$$2K(s) + 2H_2O(l) \rightarrow 2KOH(aq) + H_2(g).$$

Since the new products, KOH and H_2, have different chemical properties from the reactants, the above change is a chemical change. In addition, since the reactants cannot be formed back from the products simply by the change of the reaction conditions, such as temperature and pressure, the reaction is an irreversible one.

Do you know?

— Potassium is a metal and the fact that hydrogen gas is formed when potassium reacts with water, indicates that water is acting as an acid.
— An acid is characterized by any one of the following three possible reactions:
  • An acid reacts with a metal to give hydrogen gas.
  • An acid reacts with a carbonate/hydrogencarbonate to give carbon dioxide gas.
  • An acid reacts with a base to give salt and water.

(b) Salt dissolves in water.

Explanation:

Salt dissolves in water to give a salt solution. There is no change in the chemical composition of the salt and the water. When heat is applied to the salt solution to drive away the water, the original salt can be recovered. Thus, the dissolving of the salt in water is just a physical change.

But when the salt is dissolved, it disappears. So, shouldn't the change be a chemical one?

A: In a chemical change, the chemical property of the substance will change. Solid salt, such NaCl, consists of sodium (Na^+) and chloride (Cl^-) ions being attracted to each other. When the salt dissolves, there is no change in the chemical composition except that the ions are simply separated by the water molecules. That is, the Na^+ and Cl^- ions of the salt are not transformed into any other species that are different from themselves. During evaporation, the water molecules that are between the ions are removed and this process causes the ions to be closer to each other once again. Hence, the dissolution of salt is a physical change!

Do you know?

— The salt that dissolves in water is known as the solute, while the water is known as the solvent.

— The salt solution is homogeneous in nature as you cannot differentiate the salt from the water and the solution does not resemble the solid salt at all. Thus, these may mean that the salt solution is a compound. But we are able dissolve as much salt as possible until it does not dissolve any more. This would mean that the composition of the salt solution is variable. From this, we can conclude that the salt solution is a mixture. In addition, when the solution is evaporated to dryness, we get back the same old solid salt. Therefore, all these phenomena mean that the dissolution of salt involves physical changes and not a chemical one!

— A *mixture* is a substance that contains two or more substances which are physically together but have not chemically reacted with one another. A mixture can be a mixing of more than two elements, a mixing of more than two compounds, or a mixing of elements and compounds.

— A *compound* is a pure substance, containing two or more elements, *chemically combined* together.

— An *element* is defined as a substance which cannot, by known chemical means, be split up into two or more simpler substances.

(c) Burning a piece of paper in air.

Explanation:

Burning a piece of paper in air will result in the formation of carbon dioxide gas and water vapor. Since the new products, CO_2 and H_2O, which have chemical properties different from the reactants, the above change is a chemical change. In addition, since the reactants cannot be formed back simply by the change of the reaction conditions, such as temperature and pressure, the reaction is an irreversible one.

> (d) Heating of ammonium chloride.

Explanation:

When ammonium chloride is heated, it decomposes to form the ammonia and hydrogen chloride gases:

$$NH_4Cl(s) \rightarrow NH_3(g) + HCl(g).$$

Since the new products, NH_3 and HCl, which have chemical properties different from the reactants, the above change is a chemical change.

 But NH_3 and HCl can easily form back to the solid NH_4Cl upon cooling. So, shouldn't the above change be a physical change instead?

A: You are right that the decomposition of NH_4Cl is a reversible one. But because NH_3 and HCl are chemically different from NH_4Cl, the above change cannot be classified as a physical change. So, from this example, it is important to note that the important criteria to classify a change as a chemical one is whether new compounds of different chemical properties are formed. The reversibility of the change is a secondary criteria.

Do you know?

— The formation of NH_4Cl from $NH_3(g)$ and $HCl(g)$ through diffusion is a very good experiment to support the theory that matter is made up of small particles, such as atom, ion or molecule.

— Diffusion refers to the process that explains the movement of particles from a region of *high concentration* to one of a *lower concentration*.

cotton wool soaked with concentrated hydrochloric acid

white solid ring of ammonium chloride formed nearer to HCl

cotton wool soaked with concentrated aqueous ammonia solution

— The fact that particles of one matter can diffuse through another matter, i.e., the $NH_3(g)$ and $HCl(g)$ particles moving through the air particles, is hard evidence that there are gaps between particles in a matter. Thus, the bigger the gap between the particles, the greater the rate of diffusion. We would expect diffusion to be slow in solid, faster in liquid, and fastest in gas. In a nutshell, *diffusion through higher-density matter is slower than through lower-density matter*.

— Diffusion is dependent on the concentration gradient, i.e., the higher the concentration of the particles, the greater the rate of diffusion.

— Another factor that affects diffusion is temperature. The higher the temperature means that the particles have higher kinetic energies, resulting in a higher rate of diffusion.

— Another obvious variable is the *mass of the particle*. The larger the mass, the slower the movement of the particle, hence the lower the rate of diffusion. Thus, we would expect a denser gas, which has a higher density because of a greater particulate mass, to diffuse slower than a gas with lighter density. Therefore, the fact that the white solid ring is formed closer to the HCl is an evident that a HCl particle is heavier in mass than a NH_3 particle.

Q But doesn't a greater mass certainly mean higher K.E. since $K.E. = \frac{1}{2}mv^2$?

A: There is a fallacy here. If two particles of different masses have the same speed, then the one that has a greater mass would have a greater K.E. But if these two particles of different masses have the same K.E., then it can only mean that the one that has a greater mass must have a lower speed.

(e) Addition of water to lime juice.

Explanation:

Addition of water to lime juice does not cause the formation of new matter. Hence, the change is a physical change. In fact, this is simply a dilution process.

Do you know?

— When you add more solvent to a solution, the amount of solute does not change BUT the volume of the solution increases. So, we say that the amount of solute is "conserved"!

— Since concentration is either defined as the molar concentration (mol dm^{-3}) or mass concentration (g dm^{-3}), after dilution, the concentration of the solution decreases because ONLY the volume of the solution is changed and not the amount of substances that is dissolved in the solution.

2. The melting and boiling points of six substances are given in the following table.

Substance	Melting point/°C	Boiling point/°C
A	97	891
B	44	282
C	−40	359
D	116	184
E	−834	1432
F	−189	−185

(a) Which substances are liquids at room temperature?

Explanation:

Room temperature is about 25°C. In order for a substance to remain as a liquid at a particular temperature, the boiling point of the substance must be above this particular temperature. Similarly, in order for the substance to be a liquid, it must have a melting point that is below this particular temperature. In this case here, we must look for substances with boiling points that are above 25°C and melting points that are below 25°C. Hence, substances **C** and **E** are liquids at room temperature.

 Q In actuality, why can't a substance remain as a solid at 25°C if its melting point is below 25°C?

A: When a temperature is above the melting point or the boiling point of a substance, this means that the substance can absorb heat energy from the surroundings and undergo phase transition or change of state. But if the surrounding temperature is lower than the melting or boiling point of the substance, there is not enough "supply" of heat energy from the surroundings to "help" the substance undergo phase transition. Take for instance, a piece of ice at 0°C is left in the open with a temperature of 25°C; there is more than enough heat energy in the surroundings for the ice to absorb and then melt. But if this piece of ice is placed in a condition of 0°C, the piece of ice would not melt completely as it cannot "extract" sufficient heat from the surroundings.

Q Wait a minute, you said that at 0°C, a piece of ice cannot melt completely. Does that mean that both ice and liquid water coexist at 0°C?

A: Yes! In fact, at the melting or boiling points of a substance, both physical states of the substance coexist. A lot of students assume that at 0°C, you only have solid ice, as 0°C is the freezing point of ice. But 0°C is also the melting point of ice, so you would have the coexistence of both physical states. We say that the system is at dynamic equilibrium during the phase transition:

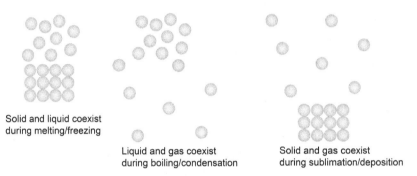

Solid and liquid coexist
during melting/freezing

Liquid and gas coexist
during boiling/condensation

Solid and gas coexist
during sublimation/deposition

Q What is "dynamic equilibrium"?

A: The term "dynamic" means change. But you would not be able to see the net change because both the rates of freezing and melting are the same. Hence, you would not be able to observe ice continue to form from the freezing of water and neither would you see water continue to form from the melting of ice. At such a state, the system is said to be at "equilibrium" as there is no net change being observed.

(b) Which substance will change its physical state when heated from 0°C to 54°C?

Explanation:

In order for a substance to change its physical state during heating, the starting temperature must be either below its melting point or boiling point or both. Since 0°C is below the melting point of substances **A**, **B**, and **D**, while 54°C is above the melting point of substance **B**, this means that only substance **B** would undergo a change of state when it is heated from 0°C to 54°C.

Do you know?

— The energy content of a solid is lower than that of the liquid, while the liquid is lower than that of a gas:

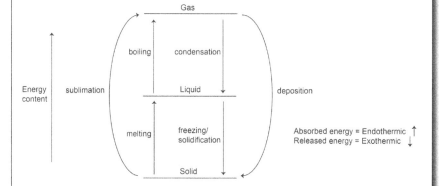

When a solid undergoes melting, the absorbed energy is used to "help" weaken the bond between the particles in the solid state. As energy cannot be created nor destroyed based on the Law of Conservation of Energy, the energy that is absorbed during melting is "transferred" as the energy content of the liquid. This thus makes the liquid possess a higher energy content than the solid. Similarly, when the liquid is converted to the gaseous state, the energy content of the gas is higher than that of the liquid.

But since energy is absorbed during a melting process, why is the measured temperature a constant value during this process?

A: The following is a heat curve that shows the corresponding changes in temperature versus the energy absorbed (in calories) as water undergoes the phase transition between the liquid and gaseous states.

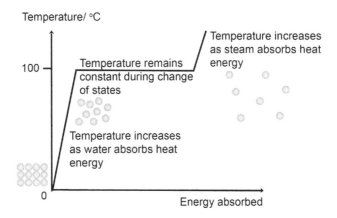

From 0°C to 100°C, the energy that the system has absorbed increases the temperature of the water because the energy that is absorbed is converted into the kinetic energy of the particles. With a higher kinetic energy, the particles move very rapidly. Due to the rapid movement, the particles cannot attract each other strongly as the distance of separation between the particles increases. At the 100°C point, the kinetic energy of the particles is sufficient to help the particles overcome the attractive forces from the other particles. Hence, the particles can escape into the gaseous phase. As a result, the energy that is introduced into the liquid will not go into increasing the temperature anymore (as the kinetic energy of the particles of liquid water no longer increases); it will be used to just send the particles of the liquid water into the gaseous state. So, imagine when the particles escape into the gaseous phase, the energy from the flame cannot be imparted onto it, so how can we have an increase in kinetic energy and in turn, an increase in temperature? Therefore, no matter how high the temperature of the flame is, a pot of boiling water will remain at 100°C until all of the liquid water has been converted to the gaseous phase. Then, further heating of the gaseous water particles without any liquid water present would then increase the kinetic energy of the gaseous particles, hence the temperature of the gas.

Q   So, does it mean that when the liquid solidifies, the same amount of energy that is absorbed during melting is given off?

A: Absolutely spot on! This is based on the Law of Conservation of Energy.

(c) Which substance exists in the liquid state over the smallest range of temperature?

Explanation:

In order to exist as a liquid, the temperature must be below the boiling point and above the melting point. Thus, to determine the range that the substance would exist as a liquid, we need to calculate the temperature range that is in between the melting and boiling points:

Substance	Melting point/°C	Boiling point/°C	Difference between boiling point and melting point/°C
A	97	891	891 − 97 = 794
B	44	282	282 − 44 = 238
C	−40	359	359 − (−40) = 399
D	116	184	184 − 116 = 68
E	−834	1432	1432 − (−834) = 2266
F	−189	−185	−185 − (−189) = 4

The substance that exists as a liquid with the smallest range of temperature is **F**.

3. Use the Kinetic Particle Theory to explain each of the following:

   (a) The volume occupied by 18 g of liquid water is 18 cm^3, but 18 g of water vapor occupies a volume of about 24,000 cm^3 at room temperature and pressure.

Explanation:

Both liquid water and water vapor consists of small particles in constant random motion. The volume of 18 g of water vapor is so much larger than the volume occupied by 18 g of water because the distance of separation between the particles in the vapor state is much larger than that in the liquid water.

 Why is the separation between the particles so much larger in the vapor state than in the liquid state?

A: The larger separation between the particles is brought about by the *weaker electrostatic attractive force* between the particles in the vapor state than in the liquid state. In a gas, the separation between particles is very large compared to their particulate sizes, such that there are virtually no attractive or repulsive forces between the particles, except during collisional contact. In a liquid, the particles are still far apart, but now they are close enough such that attractive forces confine the matter to the shape of the container that it occupies. In a solid, the particles are so close together that the forces of attraction confine the matter to a specific shape with a distinct boundary.

 So, the distance of separation between the particles is a consequence of the strength of the electrostatic attractive force and not the other way round?

A: It depends! If you have a solid, a liquid, and a gas, then based on their physical states which already are fixed, then you can say that the distance of separation between the particles is a consequence of the strength of the electrostatic attractive force. But if you are melting a solid or boiling a liquid, then when the particles absorb energy which results in an increase in the kinetic energy, the higher kinetic energy would cause the particles to move faster, hence increasing the distance of separation. This increase in the distance of separation then causes the weakening of the electrostatic force of attraction between the particles.

Q I see, so the difference in applying the concept depends on whether we are talking about the product or the process?

A: Yes! In the product (a solid, a liquid, or a gas), the distance of separation is already a consequence. But in the process, melting or boiling, the distance of separation is not a fixed consequence, but rather, it is a change because of the continual absorption of energy. This change would then lead to the consequence, i.e., a weaker attractive force.

Do you know?

— There were a few simple assumptions that were made when deriving the Kinetic Theory of Matter:

- Matter consists of a large number of small particles, which can be atoms, ions, or molecules.
- There is a large separation between these particles, be it in the solid, liquid, or gaseous state. As such, the size of the particle is *negligible* as compared to the distance of separation between the particles. Hence, the particles can be considered as *point mass*.
- The particles are in constant motion.
- As a consequence of constant motion, the particles possess *kinetic energy*, which is the energy of motion. The faster the speed of the particle, the higher the kinetic energy.
- The kinetic energy is transferred between the particles during their collisions or onto the wall of the container. There is no loss in the total energy of the system in accordance to the Law of Conservation of Energy.
- The higher the temperature of the matter, the higher the kinetic energy of the particles in the matter.
- The collision on the wall of container gives rise to the concept of pressure.

— In a nutshell, the Kinetic Theory Model assumes that matter is made up of a *large number of particles*, *widely separated*, and in *constant motion*, thus *possessing kinetic energy* which is *transferred during collision*. This model is very useful to help us understand how two or more types of matter react to give another substance. Basically, in a *chemical reaction*, different particles from different types of matter must *collide* with each other when they react and in this process, *energy transfer* takes place causing *chemical bonds* to break and form.

(b) Water vapor at 100°C can burn our skin more badly than boiling water.

Explanation:

Water vapor at 100°C can burn our skin more badly than boiling water because the water molecules in water vapor carry more kinetic energy than those in boiling water. As a result of the greater amount of kinetic energy, when the gaseous water molecules hit the skin, more heat energy is transferred to the skin, causing the skin to burn more badly.

 Is the average kinetic energy directly proportional to temperature (K.E.$_{ave}$ $\propto T$)? If so, then shouldn't water vapor at 100°C and boiling water have about the same amount of kinetic energy?

A: Yes, average kinetic energy is in fact directly proportional to temperature. But this does not mean that water vapor at 100°C has the same amount of energy as that of the boiling water. Why? This is because before the water molecules in boiling water is converted to water vapor, there is an additional amount of energy needed. This energy is known as the latent heat of vaporization, which does not lead to an increase in the temperature of the water vapor nor the boiling water. As a result, the water molecules in water vapor actually carry more energy than those in boiling water.

(c) Water boils at a lower temperature high up in a mountain than it does at sea level.

Explanation:

When water molecules gain more kinetic energy during heating, more water molecules can evaporate off from the surface. But soon, these gaseous water molecules can be "knocked" back into the liquid water by air molecules. As the temperature reaches the boiling point, more water molecules evaporate. As a result, the number of water molecules that are being "knocked" back is relatively smaller than those that have evaporated.

Hence, you have boiling taking place. Up in a mountain, the pressure is lower, meaning there are fewer air particles. As a result, we do not need a higher temperature to create a substantial amount of water molecules to "push against" the lower atmospheric pressure. Thus, the boiling point is lower.

Do you know?

— Pressure is the collisional force exerted onto the wall of the container. Mathematically, Pressure = $\frac{\text{Force}}{\text{Area}}$, whereas Force = mass × acceleration ($F = m \times a$). The amount of collisional force acting on the wall depends on the number of gas particles in the system, the volume of the system, the mass of the gas particle, and the speed of the gas particle.

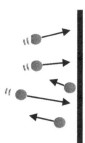

Particles colliding onto the walls of the container

— More gas particles (n) at the *same* volume and temperature as compared to another which contains fewer gas particles, would mean that there is greater collisional frequency. Hence, there is higher pressure ($p \propto n$)!

— For the same amount of gas particles, a smaller volume (V) at the *same* temperature as compared to another one of larger volume, would also mean that there is greater collisional frequency as the particles have less space to move before knocking onto the wall of container *again*. Hence, there is higher pressure ($p \propto 1/V$)!

— For the same amount of gas particles, a higher temperature (T) with the *same* volume as compared to another one of lower temperature would mean that there is greater collisional frequency as the particles have more kinetic energy and the particles move very fast. Therefore, the duration between two collisions is shorter, resulting in higher pressure ($p \propto T$)!

Why do we need a pressure cooker to cook at a higher altitude? How does a pressure cooker work?

A: As the boiling point of water is low at high altitude, this would mean that a smaller amount of heat is needed for the water to reach the lower boiling point. This lower heat content is not sufficient to cook the food. In a pressure cooker, the steam that is formed is not allowed to escape as it is sealed. As a result, pressure would slowly build up in the cooker. This built-up pressure would in turn increase the boiling point of water, hence allowing the water to take in more heat energy before it boils. With more heat content, the food can be cooked!

Q How does a pressure cooker shorten the cooking time of food?

A: The boiling point of water can be increased far beyond the 100°C than at normal atmospheric pressure. As a result, more heat energy can be transferred into the system without losing this heat energy to vaporize the liquid water. At a higher temperature, the rate of cooking is greater, thus a shorter time is needed.

4. A 100-cm-long tube was clamped horizontally as shown on p. 7. A piece of cotton wool soaked with concentrated sulfur dioxide solution was placed at one end of the tube, while another piece of cotton wool soaked in concentrated hydrogen sulfide solution was placed at the other end. The relative molecular masses of hydrogen sulfide and sulfur dioxide are 34 and 64, respectively.

   After several minutes, a ring of pale yellow solid was formed inside the tube. The word equation for the reaction is

   hydrogen sulfide + sulfur dioxide → sulfur + water.

   (a) Identify the pale yellow solid formed in the tube.

Explanation:

The pale yellow solid is sulfur.

(b) Did the yellow solid form closer to hydrogen sulfide or sulfur dioxide? Explain your answer based on the Kinetic Theory.

Explanation:

Since the rate of diffusion decreases with an increase in the mass of the particle when temperature is the same for the particles of different masses, we would expect sulfur dioxide (relative molecular mass of 64) to move slower than hydrogen sulfide. As a result, the sulfur solid would be formed closer to the sulfur dioxide.

 Q But doesn't a higher mass certainly mean higher K.E. since $K.E. = \frac{1}{2}mv^2$?

A: There is a fallacy here. If two particles of different masses have the same speed, then the one that has a greater mass would have a greater K.E. But if these two particles of different masses have the same K.E., then it can only mean that the one that has a greater mass must have a lower speed. Now, since the temperatures of the two gases are the same, this would mean that the average kinetic energies of the two gases must be the same. Hence, the heavier sulfur dioxide would have a lower speed than the lighter hydrogen sulfide.

CHAPTER 2

ATOMIC STRUCTURE

1. The elements in this table are found in Group 2 of the Periodic Table.

Element	Symbol	Electronic configuration
Beryllium	Be	2.2
Magnesium	Mg	2.8.2
Calcium	Ca	2.8.8.2
Strontium	Sr	2.8.18.8.2
Barium	Ba	2.8.18.18.8.2

(a) Which of the above elements loses electrons most readily? Explain your answer.

Explanation:

When a neutral atom loses an electron, it is the valence electron that is removed first. This is because the valence electrons are the least strongly attracted by the nucleus as compared to the other electrons that reside in the electronic shells that are before the valence shell. The following table shows the net electrostatic attractive force that is acting on the valence electrons:

Element	No. of protons (NC)	Electronic configuration	Net electrostatic attractive force acting on valence electrons (NC – shielding effect)
Beryllium	4	2.2	4 – 2 = +2
Magnesium	12	2.8.2	12 – 10 = +2
Calcium	20	2.8.8.2	20 – 18 = +2
Strontium	38	2.8.18.8.2	38 – 36 = +2
Barium	56	2.8.18.18.8.2	56 – 54 = +2

For the valence electrons, other than the attractive force that is exerted by the nuclear charge (NC), there is also inter-electronic repulsion (known as shielding effect) exerted by the inner-core electrons, which "pushes" the valence electrons away from the nucleus. Thus, the strength of the net electrostatic attractive force that is acting on the valence electrons diminishes for an atom as one progresses farther away from the nucleus.

Hence, based on the above information, you are going to expect a relatively similar amount of net electrostatic attractive force acting on the valence electrons for the above series of elements. But unfortunately, there is an additional factor that would cause a decrease in the strength of the net electrostatic attractive force that is acting on the valence electrons. This factor is the distance from the nucleus. As the distance increases from the nucleus, the net electrostatic attractive force diminishes in strength. Thus, based on this factor, we can expect the barium atom to lose its valence electrons most readily as compared to the rest in the above series.

Do you know?

— The energy level of an electronic shell that is closer to the nucleus is *lower* in energy as compared to one that is farther away.

— Bohr proposed that only a fixed number of electrons can be accommodated in any one of the energy levels. This fixed number can be calculated by using the formula $2n^2$, where n is the numbering of the energy level.

— Electrons in different energy levels are subjected to different amounts of electrostatic attractive force exerted by the nucleus. The valence shell experiences the weakest electrostatic attractive force.

— Across a period of elements in the Periodic Table, the net electrostatic attractive force that is exerted on the valence shell increases. This means that it is more difficult for the atom to lose electrons, but easier for the atom to gain electrons for elements that are on the right-hand side of a period, and vice versa for elements that are on the left-hand side.

(Continued)

(Continued)

— Down a group of elements in the Periodic Table, the net electrostatic attractive force that is exerted on the valence shell decreases because of increasing distance away from the nucleus. This means that it is easier to lose electrons but more difficult to gain electrons for elements at the bottom of a group.

— Thus, how readily the electrons can "move out" of the atom would determine the *chemical reactivity* or *chemical property* of the atom. In addition, different levels of "readiness" for the electron to be removed would also affect the overall *energy change* of the chemical reaction. Therefore, we can safely say that the chemical reactivity of an atom is *dependent* on the number of electrons and protons but independent of the number of neutrons.

— In addition, the increase of the net electrostatic attractive force across a period of elements also means that the atomic size decreases across the period. Further, a decrease of the net electrostatic attractive force down a group would mean that the atomic radius increases down a group. At any rate, since the number of electronic shells increases down a group of elements, the atomic size would naturally increase!

 Q Why does the energy level of an electronic shell that is closer to the nucleus has a *lower* energy as compared to one that is farther away?

A: By convention, when an electron is "free," i.e., not subjected to any other electrostatic interactive forces (attractive or repulsive), it has an *assigned* zero energy value. This is when the electron is infinitely away from the nucleus. And now, if you want to bring an electron from a point that is closer to the nucleus to infinity, you need to do "work" against the *electrostatic attractive force* from the nucleus; you need to "break the bond" between the electron and the nucleus. Breaking bond needs energy. The energy that you put in while doing "work" is gained by this electron (energy is conserved from Law of Conservation of Energy), hence its energy has inceased. Similarly, when an electron moves from infinity to the point that it is being attracted by the nucleus, a "bond" is formed, and thus energy is released. The following diagram would help you to understand the above explanation.

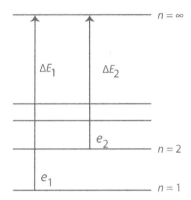

At an infinite distance from the nucleus, the energy of a free electron is zero, i.e., $E_{\text{infinity}} = 0$.

$n = 2$ has an energy level E_2, which by convention is a negative value. For example, -300 kJ mol^{-1}.

$n = 1$ has an energy level E_1, which by convention is a negative value. For example, -500 kJ mol^{-1}.

Energy of electron, e_1 (-500 kJ mol^{-1}) < Energy of electron, e_2 (-300 kJ mol^{-1}).

Energy to remove e_1, $\Delta E_1 = E_{\text{final}} - E_{\text{initial}} = 0 - E_1 = +500$ kJ mol^{-1}.

Energy to remove e_2, $\Delta E_2 = E_{\text{final}} - E_{\text{initial}} = 0 - E_2 = +300$ kJ mol^{-1}.

Since $\Delta E_2 < \Delta E_1$, we say that more energy is absorbed by Electron 1 than by Electron 2 in order to reach the same infinite distance from the nucleus. Therefore, Electron 1 must be at a lower energy level than Electron 2. Such a consideration is important in order to be in line with the concept that when energy is being absorbed, it is a positive quantity or endothermic in nature. This corresponds with "work" being done against an opposing force and in this case, it is a bond-breaking process. But during the bond-forming process, energy is released, which corresponds to a negative value or it is exothermic in nature. This thus explains why *by convention*, scientists assign a negative sign to the energy level that an electron occupies *in an atom*.

(b) What is the proton number of strontium?

Explanation:

Strontium has 38 protons.

Do you know?

— There are three fundamental sub-atomic particles in matter: **protons**, **neutrons**, and **electrons**. The table below shows the properties of these sub-atomic particles.

Properties	Proton	Neutron	Electron
Actual mass	1.673×10^{-27} kg	1.675×10^{-27} kg	9.109×10^{-31} kg
Relative mass	1	1	$\frac{1}{1840}$
Charge*	$+1.602 \times 10^{-19}$ C	0	-1.602×10^{-19} C
Relative charge	$+1$	0	-1
Location within atom	in the nucleus	in the nucleus	around the nucleus

* The SI unit used to represent the quantity of electrical charge is coulomb, C. One coulomb corresponds to one ampere $(A = C\,s^{-1})$ of electrical current flowing in one second (s).

— The protons and neutrons are collectively known as *nucleons*. The nucleons reside in the small nucleus of the atom, whereas the electrons revolve around the nucleus in the vast empty space.
— As the proton is electrically positively charged while the electron is negatively charged, the attractive force between electrically oppositely charged particles are known as *electrostatic attractive force* or Coulombic force.
— The *nuclide* of an element is represented by the notation as shown below.

$$^A_Z X$$

- The atomic symbol (X) represents each element in the Periodic Table.
- Atomic Number/Proton Number (Z) gives the number of protons in the nucleus.
- Mass Number/Nucleon Number (A) gives the sum of protons and neutrons in the nucleus.

For an electrically neutral atom, the atomic number (number of protons) is equivalent to the number of electrons.

(c) Strontium has four isotopes, ^{84}Sr (0.56%), ^{86}Sr (9.86%), ^{87}Sr (7.0%), and ^{88}Sr (82.58%). Explain the term *isotope* and calculate the relative atomic mass of strontium.

Explanation:

An element may consist of two or more atoms which have the same number of protons, also known as the atomic number, but different number of neutrons. These atoms are known as isotopes.

Relative *atomic* mass is the *average* mass of one atom of the element relative to 1/12 of the mass of one atom of ^{12}C. The relative atomic mass (A_r) of an element is dependent on (i) whether it has more than one isotope, and (ii) the composition of the various isotopes. Hence, to determine the relative atomic mass, we need the following formula:

$$A_r = \Sigma(\text{Percentage composition} \times \text{Relative isotopic mass}).$$

Relative atomic mass of strontium

$$= \frac{84(0.56) + 86(9.86) + 87(7.0) + 88(82.58)}{(0.56 + 9.86 + 7.0 + 82.5)} = 87.7.$$

Do you know?

— Different isotopes of the same element have the same chemical properties as they have an identical electronic configuration and undergo chemical reactions only involving the movement of the valence electrons. The nucleus is intact during a chemical reaction.

— Different isotopes of the same element have different physical properties such as the melting point and boiling point. Isn't it more difficult to vaporize a heavier atom from its liquid state because of its heavier mass?

(Continued)

(Continued)

— Relative *isotopic* mass is the mass of one atom of the isotope of an element relative to 1/12 of the mass of *one atom* of ^{12}C. The relative isotopic mass is almost equivalent to the relative masses of all the nucleons because the relative mass of the electrons are insignificant as compared to that of a nucleon.

— Since the dimensionless (no unit) relative isotopic mass is used in the calculation of the relative atomic mass, the latter is also a dimensionless quantity. In any rate, it is a dimensionless quantity simply because it is a relative comparison to another quantity of the same dimension (unit), which in this case, is mass.

— Based on the formula that is used for the calculation, the relative atomic mass is a *weighted average* quantity.

— The greater the contribution of a particular isotope for an element, the closer the relative atomic mass is to the value of the relative isotopic mass for the element.

— The relative composition of the isotopes for the element would be the same as that present in the compound. For example, you would find 0.56% of ^{84}Sr, 9.86% of ^{86}Sr, 7.0% of ^{87}Sr, and 82.58% of ^{88}Sr in a compound that contains one strontium atom.

— Other than the relative isotopic mass and relative atomic mass, we also have the following relative masses for molecular and ionic compounds, respectively:

- Relative *molecular* mass is the mass of one molecule of the substance relative to 1/12 of the mass of *one atom* of ^{12}C.
- Relative *formula* mass is the mass of one formula unit of the ionic compound relative to 1/12 of the mass of *one atom* of ^{12}C.

(d) If barium contains 81 neutrons, what is its nucleon number?

Explanation:

The protons and neutrons are collectively known as *nucleons*. Hence, barium has a total of (81 + 56) = 137 nucleons.

2. The number of protons, neutrons, and electrons in particles **A** to **F** are given in the following table:

Particle	Number of protons	Number of neutrons	Number of electrons
A	3	6	2
B	8	9	10
C	12	12	12
D	17	18	17
E	17	20	17
F	18	22	18

Identify which of the above particles is: (a) an atom of a metal, (b) an atom of a non-metal, (c) an atom of a noble gas, (d) a pair of isotopes, (e) a positive ion, and (f) a negative ion.

Explanation:

Particle	Number of protons	Number of neutrons	Number of electrons	Charge	Electronic configuration
A	3	6	2	+1	2
B	8	9	10	−2	2.8
C	12	12	12	0	2.8.2
D	17	18	17	0	2.8.7
E	17	20	17	0	2.8.7
F	18	22	18	0	2.8.8

(a) An atom of a metal: **C**. (Metals are *usually* from Groups 1, 2 and 13.)

(b) An atom of a non-metal: **D** and **E**. (Non-metals are *usually* from Groups 15 to 17.)

(c) An atom of a noble gas: **F**. (Group 18 elements are also known as noble gas elements.)

(d) A pair of isotopes: **D** and **E**. (Isotopes have the same number of protons but different number of neutrons.)
(e) A positive ion: **A**. (A positive ion has more protons than electrons. It is also known as a cation.)
(f) A negative ion: **B**. (A negative ion has more electrons than protons. It is also known as an anion.)

Q   What is the meaning of "noble gas"?

A: There are some elements that have low reactivity or they are relatively chemically inert, hence the term "noble gas."

Q   So, does it mean that they don't react at all?

A: Of course they do! Elements below Group 18 do react to form compounds, such as KrF_2 and XeF_4.

Do you know?

— The (Group number − 10) for an element corresponds to the number of valence electrons in the highest energy electronic shell or the outer-most electronic shell, known as the valence shell for elements that come from Groups 13 to 18.
— The period number for an element indicates the number of electronic shells that contains electrons.
— As you move across a period of elements from left to right, the elements on the left-hand side are metals, while those on the right-hand side are non-metals.
— Groups 1, 2 and 13 consist of metals; Group 14 are metalloids, which have properties between those of metals and non-metals; Groups 15 to 17 are non-metals; and Group 18 consists of the noble gases.

(Continued)

(Continued)

— Metals are characterized by their "willingness" to lose electrons or "unwillingness" to gain electrons. This is actually a result of the weak net electrostatic attractive force acting on the valence electrons.

— Non-metals are characterized by their "unwillingness" to lose electrons or "willingness" to gain electrons. This is actually a result of the strong net electrostatic attractive force acting on the valence electrons.

— As we go down any group of elements, the "willingness" to lose electrons and "unwillingness" to gain electrons increase. This is because the valence electrons are increasingly farther away from the nucleus. Hence, the valence electrons are increasingly less strongly attracted by the nucleus.

— The increase in the strength of the net electrostatic attractive force acting on the valence electrons across a period also leads to an increase in the electronegativity of the elements.

Q What is electronegativity?

A: Electronegativity refers to the ability of an atom to attract the *shared electrons* toward itself. The difference in the electronegativity values for two different elements arises because of the variation in the strengths of the net electrostatic attractive force acting on the valence shell.

Q So, since the strength of the net electrostatic attractive force increases across the period but decreases down a group, does it mean that electronegativity also increases across the period and decreases down the group?

A: Yes! Absolutely right. In addition, when electronegativity increases, electropositivity decreases. Electropositivity reflects the ability of the atom to lose electrons.

3. (a) If an element **A** forms $\mathbf{A}^{2-}$ ions,

    (i) To which group of the Periodic Table does element **A** belong?

Explanation:

If we use the octet rule as a guideline, then element **A** is likely to belong to Group 16, which has already got 6 valence electrons. Hence, element **A** needs 2 extra electrons in order to achieve the octet configuration.

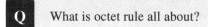 **Q**   What is octet rule all about?

A: Basically, the octet rule states that "atoms tend to lose, gain, or share electrons until they are surrounded by eight valence electrons." The octet rule stems from the fact that all noble gases (except helium) have eight valence electrons. They have very stable electronic arrangements, as evidenced by their high ionization energy, low affinity for additional electrons, and general lack of reactivity.

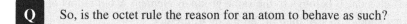 **Q**   So, is the octet rule the reason for an atom to behave as such?

A: No! Losing, gaining, or sharing of electrons until an atom contains eight valence electrons is not the reason for an atom to behave as such. Rather, the reason behind the octet rule is because of energy change. If the energy change for a reaction, in which the element needs to breach the octet rule, is too demanding, then the reaction is unlikely to occur under normal conditions. For example, if the reaction involves the formation of Na^{2+}, which needs a lot of energy to remove two electrons, then the reaction is unlikely to occur. One must take note that *there are many exceptions to the octet rule* which we will see in later sections. The octet rule should not be used as a driving force to explain why a particular compound is formed! Instead, it should just be used as a *guideline* to help us decide how many electrons are to be lost or gained or shared BUT NOT the reason for why the electron is being lost or gained or shared. Bear this clearly in mind!

(ii) The A^{2-} ion contains 18 electrons. Write down the electronic configuration of A^{2-}.

Explanation:

Starting from the $n = 1$ electronic shell, the electronic configuration of A^{2-} ion is 2.8.8.

(iii) To which period of the Periodic Table does the element belong?

Explanation:

The element **A** has an electronic configuration of 2.8.6. As there are three electronic shells that contain electrons, element **A** belongs to Period 3 of the Periodic Table.

(b) If element **B** belongs to the same group as element **A** but atoms of **B** are smaller in size than the atoms of **A**, predict

  (i) The formula of the ion which **B** forms; and

Explanation:

Since element **B** is in the same group of **A**, element **B** should also have six valence electrons. Since element **B** is smaller in size than the atom of element **A**, element **B** must be above element **A** in the same group.

  Hence, based on the above two reasons, the electronic configuration of element **B** should be 2.6. Thus, the formula of the ion which **B** forms is B^{2-}.

> **Q** Why can't the electronic configuration of element **B** be 6?

A: This is because the first electronic shell can only accommodate a maximum of two electrons. Knowing that element **B** has six valence electrons, its electronic configuration should be 2.6 and not 6.

(ii) The electronic configuration of **B**.

Explanation:

The electronic configuration of **B** is 2.6. This means that element **B** belongs to Period 2 and Group 16 (Group number = number of valence electrons + 10, for elements that come from Groups 13 to 18).

4. (a) Write the electronic configuration for the following particles: nitrogen ($_7N$), aluminium ($_{13}Al$), magnesium ion ($_{12}Mg^{2+}$), and chloride ion ($_{17}Cl^-$).

Explanation:

The electronic configuration of:

nitrogen ($_7N$) is 2.5,
aluminium ($_{13}Al$) is 2.8.3,
magnesium ion ($_{12}Mg^{2+}$) is 2.8, and
chloride ion ($_{17}Cl^-$) is 2.8.8.

 Q Why can't the electronic configuration of magnesium ion ($_{12}Mg^{2+}$) be 2.6.2 or 0.8.2?

A: When a neutral atom loses electrons, the electrons that are easiest to be removed or need the least amount of energy to be removed are the valence electrons. The inner-core electrons are more strongly attracted, with electrons in the $n = 1$ electronic shell being the most strongly attracted by the nucleus. Thus, when we want to formulate the electronic configuration of a cation, the first step is to come up with the electronic configuration of the neutral atom. Then, we remove the valence electrons in accordance to the charge of the cation that is intended to be formed, i.e. from 2.8.2 to 2.8.

 Q So why did you place the extra electron in the chloride ion into the valence shell?

A: This is because the valence shell is yet to be fully filled up, whereas the rest of the electronic shells are already fully filled according to the $2n^2$ rule.

(b) The relative atomic mass of bromine, which consists of the isotopes ^{79}Br and ^{81}Br, is 80. Calculate the percentage of ^{79}Br atoms in the isotopic mixture.

Explanation:

Let the percentage composition of isotope ^{79}Br be x. Thus, the percentage composition of isotope ^{81}Br must be $(100 - x)$.

Since the relative atomic mass of bromine $= \dfrac{79(x) + 81(100 - x)}{(x + 100 - x)} = 80$

$$\Rightarrow x = 50\%.$$

(c) Phosphorous, $^{31}_{15}$P , is an element in Group 15 of the Periodic Table.

   (i) Give the number of neutrons and protons and the electronic configuration of this atom.

Explanation:

The number 31 of $^{31}_{15}P$ is the total number of protons and neutrons (mass number or nucleon number).

The number 15 of $^{31}_{15}P$ is the total number of protons (atomic number).

The number of protons equals to the number of electrons for a neutral $^{31}_{15}P$.

The number of neutron for $^{31}_{15}P$ is $(31 - 15) = 16$.

The electronic configuration of $^{31}_{15}P$ is 2.8.5.

 Q   If we use octet rule as a guideline, does that mean that $^{31}_{15}P$ is likely to form P^{3-}?

A: Yes! The electronic configuration of P^{3-} is 2.8.8.

(ii) Draw a diagram to represent an atom of phosphorus. Show clearly the number of protons and neutrons and arrangement of electrons.

Explanation:

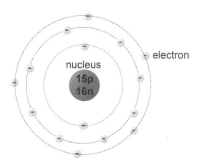

(d) Phosphorous forms an anion, P^{3-}, easily with the octet configuration. Give the number of neutrons and protons and the electronic configuration of this anion.

Explanation:

As the nucleus is still intact during a chemical reaction, the number of protons and neutrons for P^{3-} is 15 and 16, respectively. The electronic configuration of P^{3-} is 2.8.8.

> (e) Suggest two possible cations and anions that have the same electronic configuration as P^{3-}.

Explanation:

When two species have the same number of electrons, they are termed isoelectronic! Thus, the possible cations and anions that have the same electronic configuration with P^{3-} are:

The potassium ion (K^+) has an electronic configuration of 2.8.8.

The sulfide ion (S^{2-}) also has an electronic configuration of 2.8.8.

So, basically, for elements that belong to the same period as $^{31}_{15}P$, like silicon (Si), sulfur (S), and chlorine (Cl), if they gain a certain number of electrons to "follow" the octet configuration, then they would have the same electronic configuration as P^{3-}.

For elements that belong to the next period after $^{31}_{15}P$, i.e., from Period 4, if they lose electrons to "follow" the octet configuration, then they would have the same electronic configuration as P^{3-}.

Do you know?

— If two species are isoelectronic, i.e., containing the same number of electrons, this would mean that the amount of inter-electronic repulsion between the electrons is the same for the two species. So, why is it so useful to know about this?

— If both P^{3-} and S^{2-} are isoelectronic but S^{2-} has more protons than P^{3-}, this would mean that the electrons in S^{2-} are more strongly attracted than those in P^{3-}. The consequence of this is that the ionic size of S^{2-} is expected to be smaller than that of P^{3-}.

CHAPTER 3

CHEMICAL BONDING

1. (a) What do you understand by the term *element*? Compare the properties of a *mixture* to those of a *compound*. To which of these three classes of substance do the following belong: chalk, air, calcium carbonate, iodine, starch, and zinc?

Explanation:

An *element* is defined as a substance which cannot, by known *chemical means*, be split up into two or more simpler substances. A *compound* is a pure substance, containing of two or more elements *chemically combined* together. A *mixture* is a substance that contains two or more substances which are physically together but are not chemically reacted with each other.

Chalk is a mixture with the main component consisting of calcium carbonate or calcium sulfate(VI).

Air is a mixture consisting of nitrogen, oxygen, carbon dioxide, water vapor, and others.

Calcium carbonate is a compound consisting of the elements: calcium, carbon, and oxygen.

Iodine is an element.

Starch is a mixture with the main components consisting of amylose and amylopectin in varying amounts.

Zinc is an element.

Do you know?

— The following table shows the similarities and differences between an element and a compound:

Similarities between an element and a compound	Differences between an element and a compound
(a) Both are made up of particles (atoms) which are the building blocks.	(a) An element is made up of only one type of atom. A compound is made up of more than one type of atom.
(b) Both are pure substances.	(b) An element can be metallic (calcium) or non-metallic (oxygen). A compound can be formed by combining a metallic element with a non-metallic element (calcium reacts with oxygen to give calcium(II) oxide), or several non-metallic elements (nitrogen and oxygen reacts to give nitrogen dioxide).
(c) Both have fixed melting and boiling points.	(c) An element cannot be chemically split into simpler substances. A compound can be decomposed into its elements or into simpler compounds.

— The following table shows the similarities and differences between a mixture and a compound:

Similarities between a mixture and a compound	Differences between a mixture and a compound
Both are made up of more than one type of atom.	(a) A compound is homogeneous. A mixture is *usually* non-homogeneous. (b) A mixture can be separated into its components by physical methods. A compound can only be separated into its elements by chemical methods.

(Continued)

	(Continued)	
Similarities between a mixture and a compound	**Differences between a mixture and a compound**	
	(c) A mixture does not have a fixed melting or boiling point. A compound has a fixed melting and boiling point.	
	(d) No chemical changes take place when a mixture is formed. A chemical change takes place when a compound is formed.	
	(e) The proportion of the constituent components in a mixture can vary. The elements are always combined in a fixed proportion by mass in a compound.	
	(f) The properties of a mixture are the same as those of its constituent components. The properties of a compound differ from those of its constituent elements.	

(b) Which compound in part (a) has *metallic*, *ionic*, and *covalent* bonding? Explain each of this type of bonding.

Explanation:

Zinc has metallic bonding, calcium carbonate has ionic bonding, while iodine has covalent bonding.

Metallic bonding:

Within a metal, atoms partially lose their loosely bound *valence* electrons. These electrons are mobile and delocalized, not belonging to any one single atom and yet *not completely* lost from the lattice. A metal can thus be viewed as a rigid lattice of positive ions surrounded by a *sea of delocalized* electrons. What holds the lattice together is the strong metallic bonding — the electrostatic attraction between the positive ions and the delocalized valence electrons.

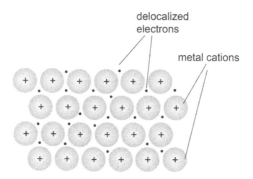

Ionic bonding:

Ionic compounds are generally formed between metals (with one, two, or three valence electrons for Groups 1, 2 and 13, respectively) and non-metals (with five, six, or seven valence electrons). The metal atoms *lose* electrons from its *outermost* valence shell, forming positively charged ions (cations). In contrast, the non-metal atoms *receive* these electrons, place them in the *outermost* valence shell, and form negatively charged ions (anions). The *resultant* electrostatic *attraction between the cations and anions is known as the ionic bond.*

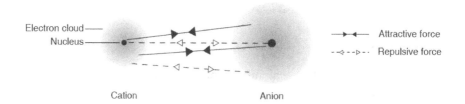

Covalent bonding:

A covalent bond is formed by the *sharing* of electrons from the valence shell between two atoms. These shared electrons are known as "bonding electrons." The shared electrons are localized between the two nuclei, within the *inter-nuclei region*, in contrast to the non-bonding electrons, which move three-dimensionally around the nucleus of its own atom. The following shows the formation of a covalent bond between two similar hydrogen atoms to form a hydrogen molecule using the electron cloud model:

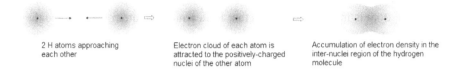

| 2 H atoms approaching each other | Electron cloud of each atom is attracted to the positively-charged nuclei of the other atom | Accumulation of electron density in the inter-nuclei region of the hydrogen molecule |

The covalent bond is the *resultant* electrostatic attraction *between* the localized shared electrons and the two positively charged nuclei.

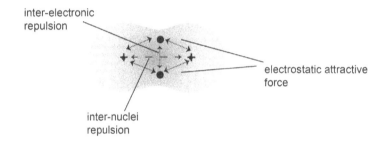

inter-electronic repulsion

electrostatic attractive force

inter-nuclei repulsion

Q So does this mean that as long as the reaction is between a metal and a non-metal, ionic bonds must be formed? But if the reaction is between two non-metals, then covalent bonds must be formed?

A: Not necessarily! A metal reacting with a non-metal to give ionic compounds is a very useful GUIDELINE, but it is not always true. For example, Al is a metal and Cl_2 is a non-metal, but $AlCl_3$ is a covalent compound. Ultimately, what makes a reaction occur is determined by whether the energy change is favorable and whether there is sufficient energy for the particles to react. Thus, we can safely say that when aluminum metal and chlorine gas react, the energy change for the reaction would "preferably" result in the formation of a covalent compound rather than an ionic compound.

What makes a metal atom more likely to lose electrons while a non-metal atom more likely to gain electrons?

A: A metal atom is more likely to lose electrons than a non-metal atom is because the net electrostatic attractive force acting on the valence shell of a metal atom is much weaker than that for a non-metal atom. Now, since the net electrostatic attractive force acting on the valence shell of a non-metal atom is much stronger, if the non-metal atom gains electrons, these extra electrons would also be strongly attracted like the rest of the valence electrons that originally belongs to the non-metal atom.

 Since the net attractive force acting on the valence shell of a non-metal atom is quite strong, does this also explain why non-metals "prefer" to form covalent bonds?

A: Yes, you are right! If two non-metal atoms "prefer" not to lose electrons, then the "best type" of bonding would involve the sharing of electrons. It is a "compromise"!

Do you know?

— Chemical bonds are electrostatic forces of attraction (a positive charge attracting a negative charge) that bind particles together to form matter. When different types of particles interact electrostatically, different types of chemical bonds are formed. There are four different types of *conventional* chemical bonds, namely, *metallic*, *ionic*, *covalent*, and *intermolecular forces*.

- Metallic bonds:
    - Strong and *non-directional*. Therefore, when a force is applied across a piece of metal, the metal atoms can slide over one

(Continued)

(Continued)

another without the breaking of the metallic bonds. This accounts for the malleability (able to be deformed into different shapes) and ductility (able to be drawn into wires) of metals.

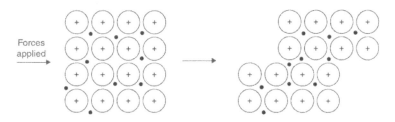

— Since metallic bonding is the result of the interaction between the delocalized electrons and the positive metal ion, the strength of a metallic bond depends on:

(i) *The number of valence electrons available for bonding.* (This factor is useful for explaining the increase in metallic bond strength across a period of metals. For example, from Na to Mg to Al, the metallic bond strength increases.)

(ii) *The size of the metal cation.* (This factor is useful for explaining the decrease in metallic bond strength down a group of metals. For example, from Na to K to Rb to CS, the metallic bond strength decreases.)

• Ionic bonds:

— Ionic compound *consists of charged cations and anions bonded together by ionic bonds*, which are difficult to break under normal conditions. Hence, when a melting or boiling process is carried out, *the heat energy is used to overcome the strong ionic bond.* This accounts for the high melting or boiling point.

— A *giant ionic lattice structure* consists of ions rigidly bonded, hence these charged particles cannot move when in the solid state. But *when melted into the molten state, these ions can act as charge carriers.* This accounts for the electrical conductivity of molten ionic compound but not for the solid compound.

(Continued)

(Continued)

— In the ionic lattice, each cation is surrounded by a number of anions and NOT just by only one anion, and vice versa. The number of cations around an anion or the number of anions around a cation is known as *coordination number*.

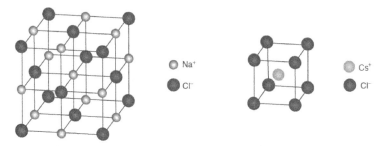

— The strength of an ionic bond depends on:

(i) *The charges of the cation and anion.* For example, the ionic bond in MgO is stronger than that in Na_2O because Mg^{2+} is doubly positively charged. Or, the ionic bond in MgO is stronger than that in MgF_2 because O^{2-} is doubly negatively charged.

(ii) *The sizes of the cation and anion.* For example, the ionic bond in MgO is stronger than that CaO because Mg^{2+} has a smaller cationic radius than Ca^{2+}, therefore it has a greater charge density $\left(\propto \frac{q_+}{r_+}\right)$. Or, the ionic bond in MgO is stronger than that in MgS because O^{2-} has a smaller anionic radius than S^{2-}.

 Q   So, the stronger ionic bond for MgO as compared to Na_2O is not related to the number of cations and anions in the formula unit (referring to the MgO or Na_2O)?

A: Of course not. The strength of ionic bond, which is measured by the melting and boiling points, is dependent on the electrostatic attractive force between a cation and an anion, instead of the number of ions that are present in the formula unit. As a cation is usually surrounded by a number of anions, and vice versa, there is more than one ionic bond formed for a cation or anion!

Is the stronger ionic bond in MgO than in Na_2O contributed by the smaller cationic size of Mg^{2+} than Na^+?

A: Certainly! Both the Mg^{2+} and Na^+ are isoelectronic (i.e., they have the same number of electrons), but since Mg^{2+} ion has a higher nuclear charge (i.e., more protons) than the Na^+ ion, the electron cloud of the Mg^{2+} ion is more strongly attracted. Hence, the distance of separation (i.e., inter-ionic separation) between the Mg^{2+} and O^{2-} ions is smaller than that between the Na^+ and O^{2-} ions. Thus, the ionic bond in MgO is stronger than that in Na_2O.

 Since the ionic bond is strong, why is an ionic compound brittle in nature?

A: Although an ionic compound is hard, because of the strong ionic bond, it is also brittle, meaning it can be broken easily. Why? This is because when a force is applied across a plane of ions, it would displace it in such a way that the cations would now face each other while the anions would also face each other. This would repel the two planes, causing the lattice to crack.

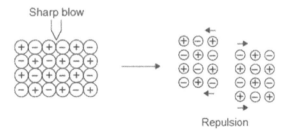

Do you know?

- Covalent bonds:

  — Covalent bond can be formed between two similar atoms of the same element or between two dissimilar atoms of different elements. Depending on the number of extra electrons each atom needs to fulfill the octet rule, there can be a single bond (e.g., H–H, Cl–Cl, H–Cl), double bond (e.g., O=O, O=C=O, C=C), or triple bond (e.g., N≡N, H–C≡C–H), which can be formed between the two atoms. Each line, "–", represents a pair of shared electrons.

  — It is also possible to have a single covalent bond being formed in which the pair of sharing electrons comes *only* from one atom, known as the *donor* atom. This type of bonding is known as *dative covalent bond* or *coordinate bond.*

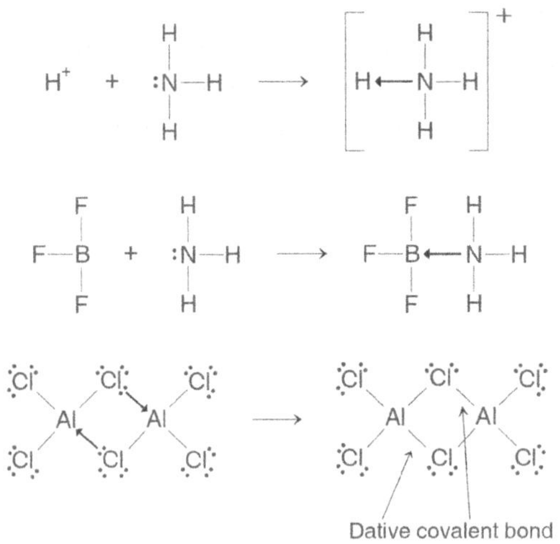

Dative covalent bond

  — Some non-metallic substances such as graphite, diamond, and silicon dioxide have high melting and boiling points which means that the attractive forces binding the particles together are very strong. For melting to occur, a great amount of energy in the form of heat

(Continued)

(Continued)

is required to overcome the strong covalent bonds *between the atoms*.

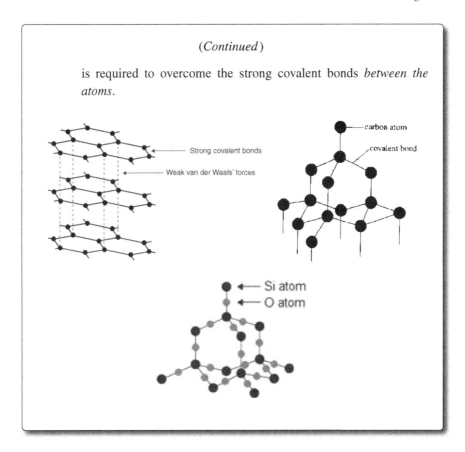

Q So all atoms that form covalent bonds would certainly have fulfilled the octet rule?

A: No! There are exceptions to the octet rule, for example:

Thus, the octet rule is not a rule that an atom must obey when the atom is involved in bond formation. Rather, it is just a useful guideline to help us predict the number of valence electrons that an atom would "prefer" to have. Ultimately, what determines the number of valence electrons that an atom would "prefer" to have during the formation of a compound is still the energy change for the reaction to form that compound.

Do you know?

- Intermolecular forces
  - Covalent bonds hold atoms in a molecule together but it is intermolecular forces that hold simple discrete molecules together. Depending on the type of molecules, there could be one of the following intermolecular forces of attractions, or a combination of these, that exist between the molecules:

    o *van der Waals' forces of attraction*

      (i) *Instantaneous dipole–induced dipole interactions (id–id)*
          Id–id interactions exist for all types of molecules but are more prominent for non-polar molecules. As electrons are always in random motion, at any point in time, there is going to have an uneven distribution of electrons in the molecule. This separation of charges creates an *instantaneous dipole* in the molecule. It is called "instantaneous dipole" because it forms instantaneously and can disappear in the next instant. The instantaneous dipole in one molecule can induce the formation of dipoles in nearby unpolarized molecules. The dipole that is being induced to form is known as *induced dipole*. As a result, a weak electrostatic attraction forms between these dipoles.

Instantaneous Induced
   dipole   dipole

          The strength of id–id interactions depends on:

          — *Number of electrons in the molecule.* The greater the number of electrons (i.e., the bigger the electron cloud),

(*Continued*)

(Continued)

the more polarizable is the electron cloud, and hence the stronger the id–id interactions.

— *Surface area for contact of the molecule.* The greater the surface area possible for contact between the molecules, the greater the extent of the id–id interactions.

(ii) *Permanent dipole–permanent dipole interactions (pd–pd)*

Pd–pd interactions exist for polar molecules only. As there is already an uneven distribution of electrons in polar bonds, permanent separation of charges (dipole) is found within the polar molecule. The permanent dipoles in neighboring polar molecules attract each other. As a result, a weak electrostatic attraction forms between these dipoles, known as pd–pd interactions.

The strength of pd–pd attractions depends on the magnitude of the molecule's *net* dipole moment. The greater the magnitude of the dipole moment, the more polar the bond.

The magnitude of the dipole, in turn, depends on the magnitude of the electronegativity difference between the bonding atoms.

The greater the difference in electronegativity between the two atoms, the greater the dipole moment being created, and thus the more polar the covalent bond.

○ *Hydrogen bonding*

Hydrogen bonding is present between molecules which have one of the highly electronegative atoms — F, O, or N — covalently bonded to a hydrogen (H) atom.

X (F, O, or N), being more electronegative than H, attracts the bonding electrons in the H–**X** bond closer to itself. As a result, **X** is more electron-rich and gains a partial negative charge ($\delta-$). The H atom, on the other hand, acquires a partial positive charge ($\delta+$) is electron deficient.

(Continued)

(Continued)

Being electron deficient, this H atom is strongly attracted to the lone pair of electrons on the highly electronegative **X** atom (electron-rich region) in other molecules. This electrostatic attraction between the H atom and the lone pair of electrons on the highly electronegative atom, F, O, or N, is known as hydrogen bonding.

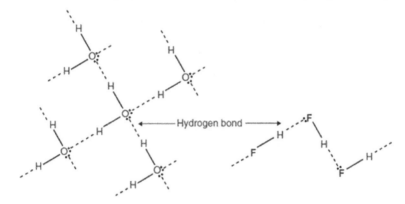

The strength of a hydrogen bond depends on:

— Dipole moment of H–**X** bond where **X** is O, F, or N

F–H- - -F–H > O–H- - -O–H > N–H- - -N–H.

— Ease of donation of the lone pair on **Y** where **Y** is O, F, or N

N–H- - -N–H > O–H- - -O–H > F–H- - -F–H.

The order of overall hydrogen bond strength is:

F–H- - -F–H > O–H- - -O–H > N–H- - -N–H.

This shows that the dipole moment is in fact the most significant factor!

It must be noted that these intermolecular attractions are also known as intermolecular bonding. They are not the same as the conventional bonds such as ionic, covalent, and metallic bonds. The attractive forces between the molecules are very *much weaker* than the conventional bonds. This accounts for why lesser amount of energy is needed to overcome intermolecular attractions than conventional bonds such as ionic, covalent, and metallic bonds. The lesser amount of energy needed is translated to lower boiling and melting points for such substances.

Q What gives rise to electronegativity?

A: Electronegativity refers to the ability of an atom to distort shared electron clouds. Now, since valence electrons are used for sharing, a more electronegative atom would have a stronger net electrostatic attractive force acting on the valence shell than a less electronegative one!

Q So, does the net electrostatic attractive force acting on the valence shell also explain why the ease of donation of the lone pair of electrons on **Y**, where **Y** is O, F, or N, decreases from left to right as shown below?

N–H- - -N–H > O–H- - -O–H > F–H- - -F–H.

A: Yes, certainly! The net electrostatic force acting on the valence shell is the strongest for a fluorine atom and weakest for a nitrogen atom. Hence, the valence electrons on the N atom are more likely to be donated than those on the O atom, which in turn is more likely to be donated as compared to those on the F atom.

Q Is the hydrogen bond the strongest type of intermolecular forces among the three?

A: It is very dangerous to make this sweeping statement, which is similar to say that an ionic bond is stronger than a covalent bond. There would be some ionic bonds that are stronger than some covalent bonds and vice versa. It is pointless to compare ionic bonds against covalent bonds just like it is useless to compare an apple against an orange.

The id–id interaction for the non-polar iodine is stronger than the hydrogen bond in water. This is why iodine is a solid at room temperature while water exists as a liquid.

(c) The elements sodium, magnesium, aluminum, silicon, phosphorous, sulfur, chlorine, and argon are the elements in Period 3. Draw simple diagrams to show the structures of three elements in Period 3.

Explanation:

Sodium, magnesium and aluminum have metallic bonding. The main differences in their metallic lattice structure would be the charge of the metal cation and the number of delocalized electrons involved in the metallic bonding:

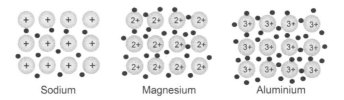

Silicon, phosphorous, sulfur, and chlorine form covalent bonds between its own atoms. Silicon has a macromolecular/giant covalent structure while phosphorous, sulfur, and chlorine all exist as simple discrete molecular substances:

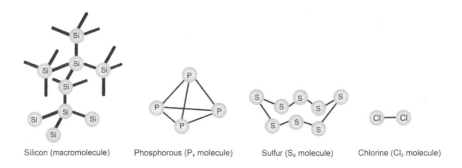

Argon exists in the monoatomic form with weak van der Waals' forces of the instantaneous dipole–induced dipole type between the atoms:

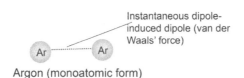

Q How would the melting points for the elements, silicon, phosphorous, sulfur, chlorine, and argon, be like? And why would they have such a trend?

A: Silicon has the highest melting point among the four elements because we are breaking Si–Si covalent bonds when we melt silicon. The melting points of sulfur, phosphorous, and chlorine would follow the trend: $S_8 > P_4 > Cl_2 > Ar$. This is because the number of electrons decreases from S_8 to P_4 to Cl_2 to Ar. The greater the number of electrons (i.e., the bigger the electron cloud), the more polarizable the electron cloud, and thus the stronger the id–id interactions. Hence, this increases the melting point.

(d) Give the formulae of (i) two covalent compounds, and (ii) two ionic compounds, formed only from elements in Period 3.
(e) Draw the dot-and-cross diagrams of the four compounds in part (d).

Explanation:

(i) Covalent compounds: $SiCl_4$, PCl_3, PCl_5 and $AlCl_3$.

$$\ddot{\underset{\cdot\cdot}{Cl}} \quad \overset{\cdot\cdot}{\underset{\cdot\cdot}{Cl}}\overset{\times}{\underset{\cdot\cdot}{Si}}\overset{\cdot\cdot}{\underset{\cdot\cdot}{Cl}} \quad \ddot{\underset{\cdot\cdot}{Cl}}$$

(ii) Ionic compounds: NaCl, $MgCl_2$ and Mg_3P_2.

$$\left[Na \right]^+ \left[:\ddot{\underset{\cdot\cdot}{Cl}}_\times \right]^- \quad \left[Mg \right]^{2+} 2\left[:\ddot{\underset{\cdot\cdot}{Cl}}_\times \right]^- \quad 3\left[Mg \right]^{2+} 2\left[{}^{\times}_{\times}\ddot{P}^{\times}_{\times} \right]^{3-}$$

Do you know?

— To draw the dot-and-cross diagram of an ionic compound:

• The metal atom loses all its valence electrons easily to form a cation. Thus, the dot-and-cross diagram for the cation should show no

(*Continued*)

(Continued)
valence electrons, but only the elemental symbol enclosed in square brackets with a superscript that denotes the charge of the cation (its value corresponding to the number of electrons lost).

- The non-metal atom "seeks" to form a stable outer octet electronic configuration. Thus, in the case of Cl, it will accept one electron. For the anion, there should be eight valence electrons depicted, and the electrons that were gained should be differentiated from the original valence electrons that the non-metal atom has. The anion would thus have a charge that corresponds to the number of electrons gained.
- The number of electrons lost by a metal atom need not necessarily be equal to that gained by a non-metal atom. To ensure electrical neutrality, you need to balance the charges by adding the required coefficients, outside the square bracket, on the left side of the symbols for the ions.

— To draw the dot-and-cross diagram of a covalent compound:

- Identify the central atom and draw the valence electrons around the central atom using either a dot or a cross symbol.
  Note: The number of valence electrons that an atom has is equivalent to its (Group number -10) for elements that come from Groups 13 to 18 in the Periodic Table.
- Draw the peripheral atoms around the central atom. Take note of the number of electrons each peripheral atom requires in order to fulfill the octet rule.
- Decide on the types of bond (dative, single, double, or triple) a peripheral atom needs to form with the central atom in order to fulfill the octet rule.

Q   Why is phosphorous able to form PCl_5? Wouldn't the phosphorous atom in PCl_5 have more than eight valence electrons?

A: Elements that cannot have more than eight electrons in the valence shell during bond formation usually come from Period 2. For these elements, the valence shell is the $n = 2$ electronic shell which can only "house" a maximum of only eight electrons. As for phosphorous, it comes from Period 3, which uses the $n = 3$ valence shell for bond formation. Since the $n = 3$ electronic shell can "house" a maximum of 18 electrons, phosphorous can form PCl_5 which breaches the octet rule.

Why is $AlCl_3$ a covalent compound but not an ionic compound?

A: Imagine when Al and Cl_2 react: if Al^{3+} and Cl^- ions are formed but due to the high charge density $\left(\propto \frac{q_+}{r_+} \right)$ of the Al^{3+} ion and the high polarizability (ability *to be* polarized) of the Cl^- ion, the electron cloud of the Cl^- would be distorted to the extent that when you observe $AlCl_3$, you would find that it is a simple molecular compound, and not an ionic one! Thus, this is again a testimony of the limitations of a scientific model. Basically, when a compound is formed from its constituent elements, you do not actually know how it is formed. It is from the already formed compound that we infer how it is formed. But is this really how it is formed? We need to do more experiments to verify it! This is also a very good example that defies the maxim, "a metal reacts with a non-metal to give an ionic compound!"

 How would the trend of the melting point of NaCl, $MgCl_2$, and Mg_3P_2 be like? And why would there be such a trend?

A: NaCl will have a lower melting point than $MgCl_2$ because the ionic bond in $MgCl_2$ is between a doubly positively charged Mg^{2+} cation and a Cl^- anion, while that in NaCl is between a singly positively charged Na^+ and a Cl^- anion. As for Mg_3P_2, it has a higher melting point than $MgCl_2$ as the ionic bond in Mg_3P_2 is between a Mg^{2+} cation and a triply negatively charged P^{3-} anion, while that in $MgCl_2$ is between a doubly positively charged Mg^{2+} cation and a singly negatively charged Cl^- anion. Hence, the ionic bond in Mg_3P_2 is stronger than in $MgCl_2$, while the ionic bond in $MgCl_2$ is stronger than that in NaCl.

 So, we cannot say that "since $MgCl_2$ has two Cl^- anions, its melting point is higher than NaCl because each Mg^{2+} is attracted to two Cl^- anions while each Na^+ is only attracted to one Cl^- anion?"

A: You cannot say that. An ionic bond is always between a cation and an anion. A cation can form more than one ionic bond in the ionic lattice, and vice versa for an anion. The number of ionic bonds that a cation or an anion can form in an ionic lattice is not determined by the chemical formula of the ionic compound. Take for instance, each Na^+ in NaCl is surrounded by six Cl^- anions, and vice versa for each Cl^- anion. This number of cations surrounding the anion is known as the coordination number. Thus, each Na^+ cation in fact forms six ionic bonds, and similarly for each Cl^- anion.

2. Sulfur is in Group 16 of the Periodic Table. There are four naturally occurring isotopes of sulfur. However, not all four isotopes are present in each sample of sulfur.

(a) Give the number of protons, neutrons, and electrons for isotopes, ^{34}S and ^{35}S.

Explanation:

Isotope	No. of protons	No. of neutrons	No. of electrons
^{34}S	16	$34 - 16 = 18$	16
^{35}S	16	$35 - 16 = 19$	16

(b) A sample of sulfur was found to contain 15% of ^{34}S and 85% of ^{35}S. Suggest how could this information be found experimentally and calculate the relative atomic mass of this sample of sulfur.

Explanation:

The isotopic composition can be determined by a method known as *mass spectrometry* (refer to *Understanding Advanced Organic and Analytical Chemistry* by K.S. Chan and J. Tan).

The relative atomic mass is a *weighted average* of the relative isotopic masses of the different isotopes:

$$A_r \text{ of sulfur} = \frac{15(34) + 85(35)}{(15 + 85)} = 34.9.$$

Did you notice that the relative atomic mass is closer to the relative isotopic mass of the isotope that has a *higher proportion* in the mixture of isotopes?

Q Why is the relative isotopic mass equivalent to the sum of the relative mass of the nucleons?

A: The bulk mass of an atom comes from the nucleus, which consists of protons and neutrons, as the mass of an electron is negligible as compared to the mass of a nucleon.

(c) The eruption of volcano releases hydrogen sulfide (H_2S), a toxic gas. Give the dot-and-cross diagram of hydrogen sulfide, showing the valence electrons only.

Explanation:

$$H \overset{\times}{\underset{\cdot\cdot}{\overset{\cdot\cdot}{S}}} \times H$$

3. Aluminum is widely used in the production of aircraft and automobiles, kitchenware, and high-voltage power cables.

(a) Describe the bonding present in solid aluminum. Explain why aluminum is a conductor of electricity.

Explanation:

The aluminum atoms are being held together by metallic bonds. The loosely bound valence electrons of each aluminum atom are delocalized. A metallic bond is the electrostatic attraction between the positive ions and the delocalized valence electrons.

As the valence electrons are delocalized, they can serve as charge carriers when an electrical potential difference is applied across a piece of metal. This thus gives rise to the electrical conductivity nature of aluminum.

Q So, when an electrical potential difference is applied across a piece of metal, the delocalized electrons actually move from one end of the metal to the other?

A: No! Although the valence electrons are delocalized, they are not so mobile that they can actually move from one end of the metal to the other. In fact, when a potential difference is applied, the electrons at one end of the metal move out of the metal first, thus creating "holes." Subsequently, nearby electrons then "hop" into these holes. This process repeated itself till the other end of the metal piece. That is really how metals conduct electricity.

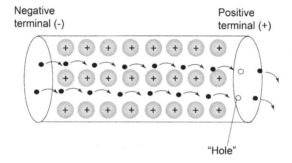

In addition, metals are also good thermal conductors due to their low mass and *mobile* delocalized valence electrons. When heat is applied to one end of the metal, the vibrational kinetic energy of the electrons and cations increases. Being low in mass, the kinetic energy of the mobile electrons will rapidly be transferred down the solid lattice structure.

(b) Give one chemical property of aluminum and support your answer with a chemical equation.

Explanation:

Being a metal, aluminum can react with acid to give hydrogen gas:

$$2Al(s) + 3H_2SO_4(aq) \rightarrow Al_2(SO_4)_3(aq) + 3H_2(g).$$

(c) Aluminum combines readily with both dry fluorine and dry chlorine. Anhydrous aluminum chloride is a white solid which sublimes at about 200°C; it reacts with water and dissolves in non-polar solvents. Aluminum fluoride melts at 1290°C and it conducts electricity; it is insoluble in non-polar solvents.

   (i) Suggest, using the information above, the name of the bond type present in both aluminum chloride and aluminum fluoride. Draw the dot-and-cross diagrams for these two compounds.

Explanation:

The useful guideline to use is, "a metal reacts with a non-metal to give an ionic compound." Based on the higher melting point of aluminum fluoride than aluminum chloride, aluminum fluoride is more likely to be an ionic compound with ionic bonding between the Al^{3+} and F^- ions. In addition, AlF_3 is insoluble in non-polar solvent, an indication of its ionic nature. In contrast, aluminum chloride is a covalent compound, with covalent bonds between the Al and the three Cl atoms. The fact that $AlCl_3$ is soluble in non-polar solvent is a good evidence of its molecular nature.

$$\left[Al\right]^{3+} 3\left[:\overset{\times}{\underset{\displaystyle \cdot\cdot}{F}}:\right]^{-} \qquad \overset{\displaystyle :\overset{\cdot\cdot}{Cl}:}{\underset{\displaystyle :\overset{\times}{\underset{\cdot\cdot}{Cl}}:}{\overset{\times}{Al}\times Cl:}}$$

Q Why are ionic compounds soluble in water but not in non-polar solvents such as hexane?

A: Polar molecules such as water molecules can interact with the ions as shown below:

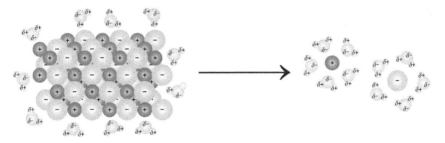

The interaction *releases* energy since such an interaction is actually a type of *bond formation* and when bonds form, energy is evolved. The energy that is released can be channeled to break the ionic bond. Non-polar solvent molecules cannot form sufficiently strong interactions with the ions. Hence, the ionic compound is insoluble in non-polar solvents as insufficient energy is released from the interaction of the ions and the solvent molecules to compensate for the energy that is needed to overcome the ionic bond. The above is the explanation behind the maxim. "like dissolves in like."

(ii) Explain why the bonding in aluminum fluoride leads to a high melting temperature, while the bonding in aluminum chloride results in a compound which sublimes.

Explanation:

The higher melting point of aluminum fluoride is due to the stronger ionic bond between the Al^{3+} and F^- ions. The lower sublimation temperature of aluminum chloride is due to the weak intermolecular forces between the $AlCl_3$ molecules.

 Q   So, the covalent bonds in $AlCl_3$ are still intact when $AlCl_3$ sublimes?

A: Absolutely right. When you melt or boil simple molecular compounds such as water, $AlCl_3$, etc., you are breaking intermolecular forces and not the *intra*molecular covalent bonds.

 Q   So, does it mean that there is still ionic bonding in a molten ionic compound?

A: Yes, of course. Similarly, there is still metallic bonding in molten metal.

So, the ability of a molten metal to conduct electricity is still due to the sea of delocalized electrons?

A: Yes!

(d) Aluminum is a preferred choice for high-voltage electric cables relative to copper. Suggest a reason for this.

Explanation:

Although copper is a better electrical conductor than aluminum due to the greater number of delocalized electrons as charge carriers, aluminum is lighter than copper. This makes aluminum able to provide a better conductivity: weight ratio than copper.

4. (a) Draw a dot-and-cross diagram to show the bonding in sodium oxide, Na_2O, and sodium chloride, NaCl.

Explanation:

$$2\left[Na\right]^{+}\left[:\overset{..}{\underset{..}{O}}\overset{x}{_x}:\right]^{2-} \quad \left[Na\right]^{+}\left[:\overset{..}{\underset{..}{Cl}}x:\right]^{-}$$

(b) Dichloromethane, CH_2Cl_2, is a covalent compound with carbon as the central atom and single bonds formed between it and the other atoms. Draw a dot-and-cross diagram for the molecule.

Explanation:

$$H\overset{\overset{\displaystyle H}{\overset{..}{x}}}{\underset{\overset{..}{\underset{..}{Cl}}}{x\,C\,x\,\overset{..}{\underset{..}{Cl}}:}}$$

(c) The following pairs of elements join together to form covalent compounds. Work out the formulae of the following compounds:

(i) silicon and chlorine;

Explanation:

Using the octet rule as a guideline, the chemical formula for silicon tetrachloride is $SiCl_4$.

(ii) carbon and oxygen;

Explanation:

Carbon and oxygen can form both CO and CO_2, in which all the atoms in both compounds fulfill the octet rule:

$$:C^{xx}_{xx}O^{x}_{x}\quad {}^{x}_{x}\ddot{O}^{xx}_{x}C^{xx}_{x}\ddot{O}^{x}_{x}$$

There is a triple bond in CO in which one of the bonds is a dative covalent bond, where the electrons are donated by the oxygen atom.

(iii) hydrogen and phosphorous;

Explanation:

Hydrogen and phosphorous can form PH_3.

(iv) carbon and chlorine;

Explanation:

Using the octet rule as a guideline, the chemical formula for carbon tetrachloride is CCl_4.

Which compound, CCl_4 or $SiCl_4$, would have a higher boiling point?

A: Since both CCl_4 and $SiCl_4$ are non-polar molecules, the intermolecular forces that hold the particles together are id–id interactions. As $SiCl_4$ has more electrons than CCl_4, its electron cloud is more polarizable. Hence, the id–id interactions between $SiCl_4$ molecules are stronger than those between the CCl_4 molecules. Thus, $SiCl_4$ would have a higher boiling point than CCl_4.

(v) carbon and sulfur; and

Explanation:

Since both sulfur and oxygen belongs to the same group, based on the octet rule, the chemical formula for carbon disulfide is CS_2.

(vi) nitrogen and fluorine.

Explanation:

Using the octet rule as a guideline, the chemical formula for nitrogen trifluoride is NF_3.

 Does NF_5 exist?

A: No. A nitrogen atom cannot have more than eight electrons in its valence shell. This is because its valence shell is the $n = 2$ electronic shell, which can only accommodate a maximum of eight electrons.

5. (a) When carbon dioxide dissolves in water, it forms carbonic acid, H_2CO_3, which contains a C=O double bond, two C–O single bonds, and two O–H single bonds. Draw a dot-and-cross diagram to represent the structure of H_2CO_3.

Explanation:

The dot and cross diagram of H_2CO_3 is:

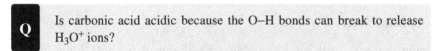

> **Q** Based on the dot-and-cross diagram, can H_2CO_3 have hydrogen bonding?

A: Yes. Since there is an H atom bonded to an O atom and there are lone pair of electrons on the various O atoms, hydrogen bonding can form between the H_2CO_3 molecules.

> **Q** Is carbonic acid acidic because the O–H bonds can break to release H_3O^+ ions?

A: Absolutely right. The H_2CO_3 can dissociate as follows:

$$H_2CO_3(aq) + 2H_2O(l) \rightleftharpoons CO_3^{2-}(aq) + 2H_3O^+(aq).$$

(b) Iodine can form either covalent or ionic bonds. Draw dot-and-cross diagrams to show the bonding in potassium iodide (KI) and iodine (I_2).

Explanation:

$$\left[K \right]^{+} \left[\ddot{\underset{..}{I}} {}_{x} \right]^{-} \qquad \ddot{\underset{..}{I}} {}_{x} \overset{xx}{\underset{xx}{I}} {}_{x}$$

(c) Explain why potassium iodide has a higher melting point than iodine.

Explanation:

The strong ionic bond in potassium iodide causes it to have a higher melting point than the intermolecular forces of the id–id type between the iodine molecules.

Do you know?

— When we melt or boil simple molecular compounds, we are breaking the intermolecular forces between the molecules and not the intramolecular covalent bonds.

(d) Explain why iodine and potassium iodide do not conduct electricity when in the solid state, but when in the molten and aqueous states, only potassium iodide can conduct electricity.

Explanation:

In the solid state, the ions in potassium iodide are rigidly held in the lattice structure. Thus, they cannot move when an electrical potential difference is applied. But when melted or in the aqueous state, both the cations and anions are mobile. Hence, these ions can act as charge carriers. As for iodine, the molecules are overall electrically neutral. Therefore, the molecules cannot move under the influence of an electrical potential difference.

(e) A boron (B) atom has one fewer electron than a carbon atom, and a nitrogen (N) atom has one more. Some B–N compounds are known to be *isoelectronic* with C–C compounds. One form of boron nitride, a colorless electrical insulator, has a planar hexagonal layered structure of alternating boron and nitrogen atoms, similar to graphite.

(i) Explain the meaning of the term *isoelectronic*.

Explanation:

When two particles have the same number of electrons, they are termed *isoelectronic*.

 Q   What is the significance of knowing the concept of being *isoelectronic*?

A: If the particles are isoelectronic, where they have the same number of electrons, the inter-electronic repulsion within each of the particles are the same.

(ii) Suggest the type of bonding which is present *within* the layers.

Explanation:

Since the structure of boron nitride is similar to that of graphite, the B and N atoms must be held together by B–N covalent bonds.

(iii) Suggest the type of interaction *between* the layers.

Explanation:

From the outset, a layer of atoms is none other than a mass of electron cloud. Since there is no covalent bond between the layers, what holds the different layers together would be intermolecular forces of the id–id type.

Do you know?

— Since the layers are being held together by intermolecular forces of the id–id dipole type, which are not strong in nature, the different layers can slide over each other easily. This allows boron nitride to be used as a lubricant, like graphite.

(iv) Suggest a possible use in which this compound would behave similarly to the corresponding carbon compound.

Explanation:

Boron nitride can replace graphite as a high-temperature lubricant.

Do you know?

— Since the atoms in boron nitride are bonded by strong B–N covalent bonds which will only break at high temperatures, it can be used as a lubricant at high temperatures.

(v) When heated under high pressure, this form of boron nitride is converted into another form which is an extremely hard solid. Suggest the type of structure adopted by this new material.

Explanation:

In this new material, each of the boron and nitrogen would adopt a tetrahedral shape. Among the four covalent bonds that an N atom forms with four B atoms, one of the N→B bond is a dative covalent bond. The overall structure of this new material is very similar to the structure of diamond.

If we have not come across the chemistry of boron nitride before, how would we know how to answer this question?

A: In this chapter on Chemical Bonding, you need to know the structures and bonding of some basic substances such as diamond, graphite, sodium chloride, water, silicon dioxide, etc. In the very beginning, the question already tells you that boron nitride is similar to graphite. So, you need to use the physical and chemical properties of graphite here. Now, since diamond is related to graphite in terms of allotropy, we thus can infer that the planar layered structure of boron nitride can be transformed into the tetrahedral form. This is applying what we have learned into a new context.

 What is the meaning of *allotropy*?

A: Allotropy refers to the phenomenon in which atoms of the same element are bonded differently together. For example, both graphite and diamond consist of carbon atoms. But in graphite, each carbon atom is bonded to three other carbon atoms. But each carbon atom is bonded to four other carbon atoms in diamond. The term allotropy is applicable only to elements but not compounds!

(f) The boiling points and molar masses of some hydrides are tabulated below.

Substance	Boiling point/K	Molar mass/g mol^{-1}
CH_4	109	16
NH_3	240	17
H_2O	373	18

(i) Suggest reasons for the difference in boiling points between NH_3 and CH_4 in terms of the type of molecules involved and the nature of the forces between them.

Explanation:

Both NH_3 and CH_4 exist as simple discrete molecular compounds held together by weak intermolecular forces. The intermolecular forces between the non-polar CH_4 molecules are of the id–id type. This is much weaker than the hydrogen bonds between the NH_3 molecules. Hence, CH_4 has a lower boiling point than NH_3.

> (ii) Why does H_2O have a higher boiling point than NH_3?

Explanation:

Both NH_3 and H_2O exist as simple discrete molecular compounds held together by hydrogen bonds. The hydrogen bonds between the H_2O molecules are more extensive than those between NH_3 molecules. This is because a H_2O molecule can form, on average, two hydrogen bonds per molecule, whereas a NH_3 molecule can only form one hydrogen bond per molecule. Hence, the boiling point of water is higher than that of ammonia.

 So, a HF molecule can also form, on average, one hydrogen bond per HF molecule?

A: Yes! This thus explains why although the hydrogen bond between two HF molecules is stronger than that between water molecules, water has a higher boiling point than HF because there are more hydrogen bonds being formed per water molecule than per HF molecule.

CHAPTER 4

MOLE CONCEPT, FORMULA, AND STOICHIOMETRY

1. Give the chemical formulae of the following:

   ammonia, ammonium nitrate, iron(III) chloride, calcium carbonate, and aluminum hydroxide.

Explanation:

Ammonia (NH_3), ammonium nitrate (NH_4NO_3), iron(III) chloride ($FeCl_3$), calcium carbonate ($CaCO_3$), and aluminum hydroxide ($Al(OH)_3$).

Do you know?

— A chemical formula represents the *proportions* of atoms that constitute a particular chemical compound. A *chemical formula* is not the name for the compound; instead, it only contains the elemental *symbols* representing the *type* of atoms and the *number* of these atoms. There are two important chemical formulae that we need to know:

  • The *empirical formula* (EF) of a compound is a representation of the *simplest* whole number ratio of the *different* atoms that made up the compound.

  • The *molecular formula* (MF) of a compound is a representation of the *actual* whole number ratio of the *different* atoms that made up the compound. Thus, MF = $n \times$ EF, i.e., an empirical formula is a *fraction* of the molecular formula.

Q Is there any difference between the chemical formula of a covalent compound and an ionic compound?

A: Certainly! The chemical formula of a simple molecular compound, such as H_2O, tells us the number and type of elements present in a molecule. But for a macromolecular/giant covalent compound, such as SiO_2, it is actually an empirical formula. It does not represent the actual number of atoms in the compound. As for an ionic compound, the chemical formula, such as KCl, is also an empirical formula, telling us the number of cations and anions present in one electrically neutral formula unit for the ionic compound.

2. Both oxygen (O_2) and ozone (O_3) are gases and they are allotropic forms of each other.
   (a) When the ozone molecule decomposes, it forms oxygen molecules. Give the balanced equation for this decomposition.

Explanation:

$$2O_3(g) \rightarrow 3O_2(g).$$

Q Why must a chemical equation be balanced? Does this mean that all the reactants have completely been converted to the products in a balanced chemical equation?

A: A balanced chemical equation does not mean that all the reactants have been converted to the products! Take for instance, two molecules of hydrogen gas (H_2) react with one molecule of oxygen gas (O_2) to produce two molecules of water (H_2O):

$$2H_2(g) + O_2(g) \rightarrow 2H_2O(l).$$

If we have three molecules of H_2 and only one molecule of O_2, there would still be one molecule of H_2 left unreacted. Why? Because the balanced equation tells us that! A chemical equation needs to be balanced simply because when matter reacts, the mass or amount of matter is the same before and after the reaction. You cannot destroy or create new matter and this is the basis for the Law of Conservation of Mass. But a balanced chemical equation does not guarantee that whatever you have put into the system would all react. Importantly, take note that matters react in specific proportions in accordance to the balanced chemical equation.

 Q   So, does it mean that O_2 is the limiting reagent here?

A: Yes, this would mean that O_2 is the limiting reagent or H_2 is the excess reagent. Hence, the actual amount of product, in this case H_2O, is dependent on the actual amount of H_2 that has reacted but not on the initial amount of H_2 that is put into the system.

Do you know?

— *Stoichiometry* shows the relative quantities of reactants and products in a chemical reaction or the relative quantities of elements in a compound. In a balanced chemical reaction, which is represented by a balanced *chemical equation*, the relationship between quantities of reactants and products typically form a *ratio* of positive integral numbers. For example, two molecules of ozone gas (O_3) decompose to produce three molecules of oxygen (O_2).

— This particular kind of stoichiometry — describing the quantitative relationships among substances as they participate in chemical reactions — is known as *reaction stoichiometry*. Stoichiometric ratio of a reactant depicting the *optimal ratio* in which the reactant is needed for the complete reaction.

— For a compound such as H_2O, the ratio of H:O = 2:1 is known as the *composition stoichiometry*. That is, it simply tells you how many atoms of each element are present in the compound. You can say that it is similar to the concept of molecular formula!

— The unit of measurement for the quantities of reactants or products in a chemical reaction is the *mole*. One mole of particles contains 6.022141×10^{23} mol^{-1} of particles.

(b) Calculate the volume of oxygen formed when $60\,cm^3$ of ozone decompose at the same temperature and pressure.

Explanation:

According to the equation: $2O_3(g) \rightarrow 3O_2(g)$, $60\,cm^3$ of O_3 would produce $\frac{3}{2} \times 60 = 90\,cm^3$ of O_2 gas.

 In a balanced chemical equation, the coefficient in front of the chemical formula indicates the amount of particles in moles, so how can we simply calculate the volume of O_2 just by using the volume of another gas?

A: The basis behind the calculation of the volume of O_2 gas is Avogadro's Law, which states that: "*equal volumes of all gases under the same conditions of temperature and pressure, contain the same number of molecules.*" With this, for a chemical equation that involves only gas particles, the chemical equation gives the mole ratio as well as the *volume ratio* of the gaseous reactants and products.

Do you know?

— The *molar volume* of one mole of any gas occupies $24.0\,dm^3$ at room temperature (20°C) and pressure (101,325 Pa or 1 atm) and $22.7\,dm^3$ at standard temperature (0°C) and pressure (10^5 Pa or 1 bar).

 Why is the molar volume of a gas smaller at a lower temperature?

A: Easy! Remember that when temperature is lower, the kinetic energy of particles is lower? At a lower K.E., the particles move slower, thus the intermolecular forces between the particles would be able to pull the particles closer together, resulting in a contraction in volume.

Q Can we talk about molar volume for a liquid?

A: No! This is because the intermolecular forces for different liquids vary in strength quite greatly. As a result, the "closeness" of the particles in the liquid state varies from liquid to liquid quite significantly. Unlike for gases, as long as the temperature and pressure are constant, the distance of separation between the gaseous particles do not vary too much.

Q So, the molar volumes for different types of gas are actually not the same?

A: Yes, in fact there is a variation. If you are interested, you can refer to *Understanding Advanced Physical Inorganic Chemistry* by J. Tan and K.S. Chan for more details.

> (c) Determine the amount of oxygen molecules in part (b) under s.t.p.

Explanation:

At s.t.p., one mole of gas occupies a volume of $22.7 \, dm^3$.

Hence, amount of oxygen molecules $= \dfrac{90}{22,700} \times 1 = 3.96 \times 10^{-3}$ mol.

 Q Can we write "mols" instead of "mol"?

A: No! "Mol" is a unit for the amount of particles just like "kg" is the unit for mass. There is no plurality for units!

> (d) Determine the volume of hydrogen gas that is needed to react with oxygen in part (c) if $4 \, cm^3$ of hydrogen reacts with $2 \, cm^3$ of oxygen. Give the balanced chemical equation based on the given information if the product formed is water.

Explanation:

If $4\,cm^3$ of hydrogen reacts with $2\,cm^3$ of oxygen, then $90\,cm^3$ of oxygen would need:

$$\frac{90}{2} \times 4 = 180 \text{ cm}^3 \text{ of hydrogen.}$$

Since $4\,cm^3$ of hydrogen contains twice the amount of particles than in $2\,cm^3$ of oxygen, the balanced chemical equation is

$$2H_2(g) + O_2(g) \rightarrow 2H_2O(l).$$

3. Determine the mass of copper that can be obtained from 65.2 g of $CuSO_4 \cdot 5H_2O$. Hence, determine the percentage by mass of oxygen in $CuSO_4 \cdot 5H_2O$.

Explanation:

The molar mass of

$$CuSO_4 \cdot 5H_2O = 63.5 + 32.1 + 4(16.0) + 5(18.0) = 249.6\,g\,mol^{-1}.$$

Amount of

$$CuSO_4 \cdot 5H_2O = \frac{65.2}{249.6} = 0.261 \text{ mol.}$$

Since there are 0.261 mol of Cu in 0.261 mol of $CuSO_4 \cdot 5H_2O$, the mass of copper that can be obtained $= 0.261 \times 63.5 = 16.6$ g.

1 mol of $CuSO_4 \cdot 5H_2O$ contains 9 mol of oxygen atoms, hence the percentage by mass of oxygen in

$$CuSO_4 \cdot 5H_2O = \frac{9(16.0)}{249.6} \times 100\% = 57.7\%.$$

Do you know?

— *Molar mass* is the mass of *one mole* of a substance and is *numerically* equal to the relative atomic mass or the relative molecular mass of the substance expressed in grams. The unit for molar mass is $g\,mol^{-1}$.

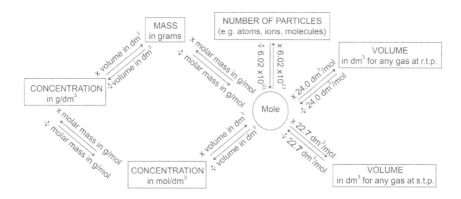

4. An unknown compound is found to have a molar mass of $194.19\,g\,mol^{-1}$. From elemental analysis, its percentage composition by mass is found to contain: 49.48% C, 5.19% H, 28.85% N, and 16.48% O. Determine both the empirical and molecular formulae of the compound.

Explanation:

Elements present	C	H	N	O
Assuming 100 g of mass	49.48	5.19	28.85	16.48
Amount (in mol)	49.48/12 = 4.12	5.19/1 = 5.19	28.85/14 = 2.06	16.48/16 = 1.03
Divide by smallest number of mole	4.00	5.04	2.00	1.00
Simplest ratio	4	5	2	1

The empirical formula of the unknown compound is $C_4H_5N_2O$.
Since molecular formula $= n \times$ empirical formula,

$$n \times (4 \times 12 + 5 \times 1 + 2 \times 14 + 16) = 194.19$$

$$n = 2.$$

Hence, the molecular formula is $C_8H_{10}N_4O_2$.

---

5. Barium carbonate reacts with nitric acid according to the equation:

$$BaCO_3(s) + 2HNO_3(aq) \rightarrow CO_2(g) + Ba(NO_3)_2(aq) + H_2O(l).$$

In an experiment, 25.0 g of barium carbonate was completely reacted.

(a) Determine the amount of carbon dioxide evolved in terms of the number of moles and its mass.

Explanation:

Molar mass of $BaCO_3 = 137.3 + 12.0 + 3(16.0) = 197.3\,g\,mol^{-1}$.

Amount of $BaCO_3$ in 25.0 g $= \dfrac{25.0}{197.3} = 0.127\,mol$.

One mole of $BaCO_3$ gives one mole of CO_2, hence the amount of CO_2 evolved $= 0.127\,mol$.

Molar mass of $CO_2 = 12.0 + 2(16.0) = 44.0\,g\,mol^{-1}$.

Mass of $CO_2 = 0.127 \times 44.0 = 5.59\,g$.

(b) What is the concentration of HNO_3 that is needed if 50 cm^3 of the acid were used to completely react with the barium carbonate?

Explanation:

One mole of $BaCO_3$ reacts with two moles of HNO_3, hence the amount of HNO_3 used $= 2 \times 0.127 = 0.254$ mol.

$50\,cm^3$ is equivalent to $\dfrac{50}{1000} = 0.05\,dm^3$.

Concentration of $HNO_3 = \dfrac{0.254}{0.05} = 5.08\,mol\,dm^{-3}$.

(c) What would be the concentration of sulfuric(VI) acid that is needed if only $32\,cm^3$ of sulfuric acid were used?

Explanation:

The reaction equation between $BaCO_3$ and H_2SO_4 is

$$BaCO_3(s) + H_2SO_4(aq) \rightarrow CO_2(g) + BaSO_4(s) + H_2O(l).$$

One mole of $BaCO_3$ reacts with one mole of H_2SO_4, hence the amount of H_2SO_4 used is 0.127 mol.

$32\,cm^3$ is equivalent to $\dfrac{32}{1000} = 0.032\,dm^3$.

Concentration of $H_2SO_4 = \dfrac{0.127}{0.032} = 3.97\,mol\,dm^{-3}$.

(d) Determine the mass of barium sulfate ($BaSO_4$) that would be completely precipitated out by the sulfuric(VI) acid in part (c).

Explanation:

One mole of $BaCO_3$ gives one mole of $BaSO_4$, hence the amount of $BaSO_4$ precipitated out is 0.127 mol.

Molar mass of $BaSO_4 = 137.3 + 32.1 + 4(16) = 233.4 \, g \, mol^{-1}$.

Mass of $BaSO_4$ precipitated out $= 0.127 \times 233.4 = 29.6 \, g$.

6. 5.77 g of white phosphorous and 5.77 g of oxygen were mixed and the following reaction to form phosphorous(III) oxide occurred:

$$P_4(s) + 3O_2(g) \rightarrow P_4O_6(s).$$

In excess oxygen, the phosphorous(III) oxide can further react to produce phosphorous(V) oxide in accordance to the following equation:

$$P_4O_6(s) + 2O_2(g) \rightarrow P_4O_{10}(s).$$

(a) Determine which reactant is the limiting reagent. Hence, calculate the mass of P_4O_6 formed.

Explanation:

Amount of P_4 in $5.77 \, g = \dfrac{5.77}{4(31.0)} = 4.65 \times 10^{-2}$ mol.

Amount of O_2 in $5.77 \, g = \dfrac{5.77}{2(16.0)} = 0.180$ mol.

One mole of P_4 reacts with three moles of O_2; 4.65×10^{-2} mol of P_4 would react with $3 \times 4.65 \times 10^{-2} = 0.1395$ mol of O_2.

P_4 is the limiting reagent!

One mole of P_4 gives one mole of P_4O_6; 4.65×10^{-2} mol of P_4 would give 4.65×10^{-2} mol of P_4O_6.

Molar mass of $P_4O_6 = 4(31.0) + 6(16.0) = 220.0 \, g \, mol^{-1}$.

Mass of P_4O_6 formed $= 4.65 \times 10^{-2} \times 220.0 = 10.23 \, g$.

Do you know?

— Phosphorous exists as simple discrete molecules, P_4, and not in the monoatomic form.

(b) Determine the volume of oxygen at r.t.p. that is needed to convert the phosphorous(III) oxide to phosphorous(V) oxide.

Explanation:

One mole of P_4O_6 reacts with two moles of O_2; 4.65×10^{-2} mol of P_4O_6 would react with $2 \times 4.65 \times 10^{-2} = 9.30 \times 10^{-2}$ mol of O_2.

At r.t.p., one mole of gas occupies volume of $24.0\,dm^3$.

Hence, volume of O_2 gas needed $= 9.30 \times 10^{-2} \times 24.0 = 2.23\,dm^3$.

(c) Calculate the mass of P_4O_{10} that is produced.

Explanation:

One mole of P_4O_6 gives one mole of P_4O_{10}; 4.65×10^{-2} mol of P_4O_6 would give 4.65×10^{-2} P_4O_{10}.

Molar mass of $P_4O_{10} = 4(31.0) + 10(16.0) = 284.0\,g\,mol^{-1}$.

Mass of P_4O_{10} produced $= 4.65 \times 10^{-2} \times 284.0 = 13.21\,g$.

(d) When P_4O_{10} dissolves in water, it forms phosphoric(V) acid, H_3PO_4, which is a tribasic acid:

$$P_4O_{10}(s) + 6H_2O(l) \rightarrow 4H_3PO_4(s).$$

Determine the number of moles of $Ca(OH)_2$ that is needed to completely react with all the H_3PO_4 formed. Hence, calculate the concentration of the $Ca(OH)_2$ solution that is needed if $50\,cm^3$ of it was used, in both the mass and molar concentrations.

Explanation:

One mole of P_4O_{10} gives four moles of H_3PO_4; 4.65×10^{-2} P_4O_{10} would give $4 \times 4.65 \times 10^{-2} = 0.186$ mol of H_3PO_4.

One mole of a tribasic acid would react with three moles of OH^- ions to produce three moles of H_2O.

Hence, two moles of H_3PO_4 would react with three moles of $Ca(OH)_2$:

$$2H_3PO_4 + 3Ca(OH)_2 \rightarrow Ca_3(PO_4)_2 + 6H_2O.$$

Amount of $Ca(OH)_2$ needed to react with 0.186 mol of

$$H_3PO_4 = \frac{3}{2} \times 0.186 = 0.279 \text{ mol}.$$

$50 \, cm^3$ is equivalent to $\dfrac{50}{1000} = 0.05 \, dm^3$.

Concentration of $Ca(OH)_2$ solution $= \dfrac{0.279}{0.05} = 5.58 \text{ mol dm}^{-3}$.

Molar mass of $Ca(OH)_2 = 40.1 + 2(16.0 + 1.0) = 74.1 \, g \, mol^{-1}$.

Concentration of $Ca(OH)_2$ solution in $g \, dm^{-3} = 5.58 \times 74.1 = 413.5 \, g \, dm^{-3}$.

CHAPTER 5

ENERGY CHANGE AND FUELS

1. The chemical equation for the addition of chlorine to ethene is given as follows:

The following is a table of bond energies for some bonds:

Covalent bond	Bond energy in kJ mol^{-1}
H–C	410
C–C	350
C=C	610
Cl–H	431
Cl–Cl	244
C–Cl	340

(a) How much energy is needed in total to break one mole of C=C bond and one mole of Cl–Cl bond?

Explanation:

Total energy that is needed to break one mole of C=C bond and one mole of Cl–Cl bond = 610 + 244 = 854 kJ mol^{-1}.

Do you know?

— Bond energy is the amount of heat energy that is *required to break* one mole of covalent bonds. This is an endothermic quantity as energy is absorbed during the bond-breaking process.
— The stronger the covalent bonds between the atoms, the more endothermic or positive would the bond energy value be.

 Where does the energy that is required to break a bond come from and where does it go to?

A: Good question. The source of energy that is required to break a bond is in the form of kinetic energy (K.E.) that the particles possess and this K.E. is transformed into the potential energy (P.E.), which is being *stored in the bond*. What will happen when the amount of K.E. is large? The particles will vibrate very fast. This would mean that the *distance* of separation between the particles would increase. The resultant increase in the distance of separation would mean that the P.E., which is being stored within the bond, will also increase. Likewise, when a bond is formed, the P.E. is converted back to K.E. Thus, it is important to remember that the *energy that is required to break a bond would be stored as P.E. in the bond*.

 So does that mean that the strength of a chemical bond is actually nature's energy storage facility?

A: Yes. The *stronger* the bond, the *lower* the amount of P.E. stored, and the *more stable* the substance. For example, carbon dioxide and water in total have a smaller amount of stored energy as compared to glucose and oxygen. During photosynthesis, plants absorb sunlight and transfer the absorbed energy as stored energy in the weaker chemical bonds in glucose and oxygen. And when these two compounds react with each other, the energy that is stored is then released to form more stable substances containing stronger bonds.

(b) (i) List the bonds that are formed in the reaction.

Explanation:

One C–C single bond and two C–Cl single bonds are formed in the reaction.

(ii) How much energy is released in forming all these bonds?

Explanation:

Energy released = 350 + 2(340) = 1030 kJ mol^{-1}.

Q Why isn't a negative sign placed in front of the 1030 kJ mol^{-1}?

A: This is because there is already the phrase, "energy released" in front of the equation, so it is understood that the 1030 kJ mol^{-1} is an exothermic quantity. But if you use the phrase "energy change" instead, then the value presented should be −1030 kJ mol^{-1}. This is because the word "change" does not indicate the direction of change, which thus needs to be indicated by either a "+" or "−" sign.

Do you know?

— Bond formation is an exothermic process, i.e., energy is released. The amount of energy that is released during bond formation is equal to the amount of energy that is needed to break the same bond.
— The stronger the bond that is formed, the more exothermic or negative the energy change.

(iii) Refer to Section 5.5 of *Understanding Basic Chemistry* by K. S. Chan and J. Tan, and calculate Δ*H* for the reaction.

Explanation:

Section 5.5 of *Understanding Basic Chemistry* by K.S. Chan and J. Tan states that:

$$\Delta H = \text{(Sum of total energy for bond breaking)}$$
$$- \text{(Sum of total energy for bond making)}.$$

Hence, $\Delta H = 854 - 1030 = -176$ kJ mol^{-1}.

Do you know?

— An exothermic reaction has a negative ΔH while an endothermic reaction has a positive ΔH:

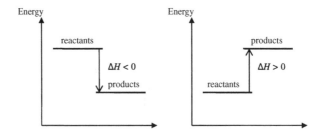

— The magnitude and sign of ΔH provide us with important information. Other than informing us of the *stability of the product relative to that of the reactant*, the ΔH value also reflects the *strength of the bonds in the product relative to those in the reactant.*

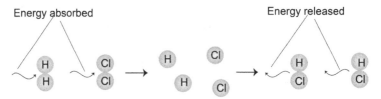

A negative ΔH (exothermic) indicates that the bonds in the product are stronger than those in the reactant, and vice versa. That is, for an exothermic re*action:*

ΔH for bond breaking in product > ΔH for bond breaking in reactant.

But for an endothermic reaction, it would be:

ΔH for bond breaking in reactant > ΔH for bond breaking in product.

(c) With reference to the given information, explain whether a double bond between two carbon atoms is twice as strong as a single bond.

Explanation:

Covalent bond	Bond energy in kJ mol^{-1}
C–C	350
C=C	610

The strength of the C=C bond is not twice that of the C–C bond! This is because if the strength of the C=C bond is twice that of the C–C bond, then the energy that is needed to break the C=C bond would be 700 kJ mol^{-1} and not 610 kJ mol^{-1}.

What is the actual reason for the bond strength of the C=C double bond not being exactly twice as strong as that of the C–C single bond?

A: A covalent bond is the resultant electrostatic attractive force that the nuclei of the two sharing atoms acting on the shared electrons. Although a double bond is stronger than a single bond because the electrostatic attractive forces that are acting on four electrons are much greater than the electrostatic attractive forces that are acting on two electrons, the inter-electronic repulsion between the four electrons is also much greater than that between the two electrons in a single bond. Hence, the greater inter-electronic repulsion between the greater number of shared electrons weakens the double bond.

(d) Draw the energy level diagram for the above reaction. Hence, explain the sign of the energy change.

Explanation:

The exothermic energy change, which is indicated by the negative sign, shows that the bonds that are formed are stronger than the bonds that have been broken during the chemical reaction. The product is thus more stable than the reactants.

Q   Why does a lower energy level confer greater stability?

A: As the enthalpy of a substance is a measurement of the energy content of a substance, it also reflects its *stability*. The greater the enthalpy, the greater the energy content, and the less stable the substance. Why? This is because if the energy content of the substance is high, this would mean that either the particles in the substance are moving fast (large amount of K.E.) or the bond between the particles is weak (large amount of P.E.). So, how can a weak bond hold particles together? How can fast-moving particles attract each other strongly?

Q   So, the less stable the substance, the more reactive it can be. Thus, is the high reactivity related to the weak bonds in the substance?

A: Certainly. The less stable the substance, the weaker the bond, the easier for the bonds to break, and hence, the more reactive the substance.

(e) If the energy required to break one mole of Br–Br bond is 193 kJ mol^{-1}, how would you expect the energy change for the addition of bromine to ethene to be like?

Explanation:

The total energy that is required to break a C=C and a Br–Br bonds is $= 610 + 193 = 803$ kJ mol^{-1}.

Assuming that the C–Br bond that is formed is of the same strength as the C–Cl bond, then

$$\Delta H = 803 - 1030 = -227 \text{ kJ mol}^{-1}$$

i.e., we may expect the energy change to be more exothermic (more negative than -176 kJ mol^{-1}) than that for the reaction between ethene and chlorine.

 Is the C–Br of the same strength as the C–Cl bond?

A: Of course not. The bond energy of a C–Br bond is 288 kJ mol^{-1}. Hence, the actual energy change should be

$$\Delta H = 803 - \{350 + 2(288)\} = -123 \text{ kJ mol}^{-1}.$$

Did you notice that the actual value is less exothermic than that for the reaction between ethene and chlorine?

 Since the actual bond energy for the C–Br bond is not given, what can we do during an examination if such a question is asked?

A: No worries, simply make an assumption and then work out the solution from there as what we have done in the question here. As long as we have demonstrated understanding of the chemistry, it is good enough!

(f) The reaction of ethene with oxygen is represented by the equation:

$$C_2H_4(g) + 3O_2(g) \rightarrow 2CO_2(g) + 2H_2O(l),$$

$$\Delta H = -1140 \text{ kJ mol}^{-1}.$$

100 cm^3 of ethene was mixed with excess oxygen and ignited. Assume that all measurements were made at 20°C and 1 atm. Calculate:

(i) The volume of oxygen that was needed for the complete combustion.

Explanation:

One mole of $C_2H_4(g)$ reacts with three moles of $O_2(g)$; based on Avogadro's Law, 100 cm^3 of $C_2H_4(g)$ will react with $3 \times 100 = 300 \text{ cm}^3$ of $O_2(g)$.

Do you know?

— Avogadro's Law states that equal volumes of all gases under the same conditions of temperature and pressure, contain the same number of particles. That is, 10 cm^3 of $C_2H_4(g)$ has the same number of particles as 10 cm^3 of $O_2(g)$ or 10 cm^3 of $CO_2(g)$.

(ii) The volume of carbon dioxide produced.

Explanation:

One mole of $C_2H_4(g)$ gives two moles of $CO_2(g)$; based on Avogadro's Law, 100 cm^3 of $C_2H_4(g)$ will give $2 \times 100 = 200 \text{ cm}^3$ of $CO_2(g)$.

(iii) The amount of heat energy evolved during the reaction.

Explanation:

At 20°C and 1 atm (r.t.p.), one mole of gas occupies a volume of $24.0\ dm^3$.

$100\ cm^3$ is equivalent to $\dfrac{100}{1000} = 0.1\ dm^3$.

The amount of $C_2H_4(g)$ in $100\ cm^3 = \dfrac{0.1}{24.0} \times 1 = 4.17 \times 10^{-3}\ mol$.

When one mole of $C_2H_4(g)$ reacts, 1140 kJ of heat energy is released. The amount of heat energy that is released for 4.17×10^{-3} mol of $C_2H_4(g)$ reacted $= 1140 \times 4.17 \times 10^{-3} = 4.75$ kJ.

(iv) Explain why the amount of heat energy measured would be less than the expected value.

Explanation:

The amount of heat energy that is measured would be less than the expected value because during the measurement process, some heat energy would have been lost to the surroundings and so, it is not "captured" during the measurement.

Q Wow! Isn't this very similar to the situation in which not all the heat energy that is released from the burning of LPG (liquefied petroleum gas) would be transferred to boil a kettle of water?

A: Absolutely right! Some of the heat energy would be lost to the surroundings.

Do you know?

— The heat change of a reaction can be measured using a *calorimeter*. A simple set-up is shown below:

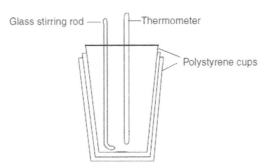

Glass stirring rod

Thermometer

Polystyrene cups

The reaction happens in the polystyrene cup, which provides thermal insulation. The glass rod is to mix the reactants well so that hopefully, all the reactants react at *one go*. The heat is evolved at *one go with minimal lost* to the surroundings. The thermometer records the initial and highest temperature rise, which give us the temperature difference. But if heat is absorbed instead, then the final temperature that is recorded would be lower than the initial temperature. This temperature difference is *proportional* to the amount of heat energy that is released or absorbed.

2. At room temperature, aqueous hydrogen peroxide decomposes very slowly to form water and oxygen. The chemical equation representing the decomposition is below:

$$2H-O-O-H \rightarrow O=O + 2H-O-H, \quad \Delta H = -206 \text{ kJ mol}^{-1}.$$

(a) Explain why this reaction is exothermic in terms of the energy changes that take place during bond-breaking and bond-making processes. Hence, draw the energy level diagram for the reaction.

Explanation:

In the reaction, two O–O single bonds are broken and energy is absorbed during the bond-breaking process. When an O=O double bond is formed, energy is released during the bond-making process. Since the strength of the O=O double bond is stronger than the sum of the two O–O single bonds, the overall energy change would be exothermic in nature as more energy is being released than absorbed.

The energy level diagram for the above reaction is as follows:

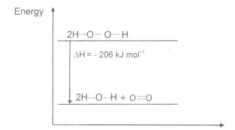

 Q   How do you actually know that "the strength of the O=O double bond is stronger than the sum of the two O–O single bonds?"

A: You have to make this deduction accordingly here, otherwise you won't be able to give a reasonable explanation as to why the overall energy change is exothermic. Remember that an exothermic energy change means that the bonds in the products are stronger than the bonds in the reactants? Likewise for an endothermic one!

Q   The hydrogen peroxide decomposes easily. Is it because of the negative enthalpy change?

A: Not all reactions with a negative enthalpy change indicate that the reactions would proceed spontaneously. Take for instance, diamond is less stable than graphite, but diamond does not spontaneously convert to graphite, because the conversion process has a high activation energy as shown:

$$C \text{ (diamond)} \rightarrow C \text{ (graphite)}, \quad \Delta H = -1.83 \text{ kJ mol}^{-1}.$$

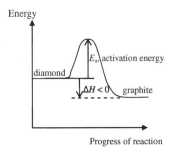

Energy

E_a activation energy

diamond

$\Delta H < 0$ graphite

Progress of reaction

Many reactions are energetically feasible (exothermic), but they only occur very slowly due to the high activation energy that is involved — the energy barrier that is needed to be surmounted before a successful reaction can occur. Such reactions are said to be kinetically non-feasible.

Do you know?

— A double bond is stronger than a single bond because the electrostatic attractive forces of two nuclei acting on the four shared electrons are stronger than if the forces of the two nuclei are acting on only two shared electrons. The same reasoning is applicable if we compare a triple bond to a double bond.

Q

According to the equation, there are two moles of H_2O_2 decomposed:

$$2H-O-O-H \rightarrow O=O + 2H-O-H, \quad \Delta H = -206 \text{ kJ mol}^{-1},$$

yet the unit for the energy change is in kJ mol^{-1}, is there something wrong here?

A: Although there are two moles of H_2O_2 in the equation for the decomposition reaction, the value of -206 kJ mol^{-1} is meant to indicate the amount of energy that is evolved for the decomposition of two moles of H_2O_2 and not for one mole of H_2O_2. If there is only one mole of H_2O_2 decomposed, then the equation would be:

$$H-O-O-H \rightarrow \frac{1}{2} O=O + H-O-H, \quad \Delta H = -103 \text{ kJ mol}^{-1}.$$

If there are three moles of H_2O_2, then the equation would become:

$$3H–O–O–H \rightarrow \frac{3}{2}O{=}O + 3H–O–H, \quad \Delta H = -309 \text{ kJ mol}^{-1}.$$

Did you notice that the unit for the equation is still kJ mol^{-1}? But the numerical values are not the same. Instead, they are proportional to the number of moles of H_2O_2 that is being stated in the equation. Thus, it is common to interpret the "per mole" in the unit, kJ mol^{-1}, as per mole of the "whole equation" and not just per mole with respect to just one mole of the reactant.

(b) Calculate the amount of energy evolved when 2 g of hydrogen peroxide decomposes.

Explanation:

Molar mass of $H_2O_2 = 2(1.0) + 2(16.0) = 34.0$ g mol^{-1}.

Amount of $H_2O_2 = \dfrac{2}{34.0} = 5.88 \times 10^{-2}$ mol.

When two moles of H_2O_2 decompose, 206 kJ of heat energy is released.
   The amount of heat energy that is released for 5.88×10^{-2} mol of H_2O_2 decomposed $= 206 \times (5.88 \times 10^{-2}/2) = 6.06$ kJ.

(c) Draw the dot-and-cross diagram of hydrogen peroxide.

Explanation:

$$H{:}\ddot{O}{:}\overset{xx}{\underset{xx}{\ddot{O}}}{:}H$$

Q There are two O–H groups in hydrogen peroxide — does it mean that the molecule is capable of forming hydrogen bonds?

A: Good observation. Yes, hydrogen peroxide molecules can form hydrogen bonds between themselves.

(d) Given the bond energies of the following:

Covalent bond	Bond energy in kJ mol^{-1}
O=O	486
O–H	463

Calculate the bond energy of O–O bond in hydrogen peroxide.

Explanation:

Let the bond energy of O–O bond in hydrogen peroxide be BE(O–O).

Total energy absorbed during bond-breaking = $2 \times$ BE(O–O).

Total energy released during bond forming = $1 \times 486 = 486$ kJ mol^{-1}.

$$\Delta H = \text{(Sum of total energy for bond breaking)}$$
$$- \text{(Sum of total energy for bond making)}.$$

Hence, $\Delta H = 2 \times \text{BE(O–O)} - 486 = -206$ kJ mol^{-1}

$$\Rightarrow 2 \times \text{BE(O–O)} = 486 - 206$$
$$\Rightarrow \text{BE(O–O)} = 140 \text{ kJ mol}^{-1}.$$

Interesting, previously we have discussed that the strength of the C=C double bond is not twice that of the C–C single bond. Yet, the strength of the O=O double bond (483 kJ mol^{-1}) is actually more than twice the strength of the O–O single bond (140 kJ mol^{-1}). Why is it so?

A: There is a different approach to account for this seemingly contradicting observation. When a carbon atom is involved in covalent bond formation to form a stable compound, there are no "unused" electrons around the carbon atom, i.e., all electrons are used up. As for oxygen, the formation of the O=O double bond results in the accumulation of four electrons within the inter-nuclei region. Yes, there is much more inter-electronic repulsion among these four shared electrons as compared to the two shared electrons in the O–O bond. But the inter-electronic repulsion between the "unused" electrons on the two oxygen atoms of the O=O double bond (four "unused" electrons on each of the two O atoms) is smaller than that in the O–O bond (five "unused" electrons on each of the O two atoms). This effect actually helps to strengthen the O=O double bond. This means that the greater inter-electronic repulsion between the "unused" electrons on the two oxygen atoms of the O–O single bond actually further weakens the bond.

3. In the future, fuel cells may be used to power cars. In a fuel cell, the overall reaction is represented by the equation

$$2H_2(g) + O_2(g) \rightarrow 2H_2O(l), \quad \Delta H = \text{negative}.$$

(a) Explain the term *fuel cell*.

Explanation:

A fuel cell is a device that uses a fuel to react with oxygen in the air to produce electrical energy *directly*. The oxidizing agent which is oxygen gas, is supplied to the cathode compartment. The fuel, which can be hydrogen gas, hydrazine ($H_2N–NH_2$), methanol (CH_3OH), sugar ($C_6H_{12}O_6$), and other organic compounds, is fed to the anode compartment.

Do you know?

— The fuel cell consists of two electrodes (made of platinum or palladium), which are in contact with the electrolyte (either a NaOH or H_2SO_4 solution). The *negative* electrode (anode, where oxidation takes place) is supplied with hydrogen (the fuel) while the *positive* electrode (cathode. where reduction takes place) is supplied with oxygen.

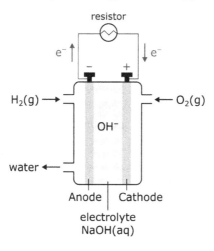

(*Continued*)

(Continued)

— The fuel undergoes oxidation at the anode, releasing electrons. The electrons flow out of the cell through the external circuit and then return to the cathode, producing an electric current along the way. At the cathode, the oxygen undergoes reduction by "consuming" the electrons.

Anode: $2H_2 + 4OH^- \rightarrow 4H_2O + 4e^-$

$H_2N-NH_2 + 4OH^- \rightarrow 4H_2O + N_2 + 4e^-$

$CH_3OH + 6OH^- \rightarrow 5H_2O + CO_2 + 6e^-$

$C_6H_{12}O_6 + 24OH^- \rightarrow 18H_2O + 6CO_2 + 24e^-.$

Cathode: $O_2 + 2H_2O + 4e^- \rightarrow 4OH^-.$

— The greater the amount of electrons that is generated in the anodic process, the greater the amount of electrical energy generated. Other than sodium hydroxide, sulfuric acid can also be used. Both electrolytes are known for causing high-temperature corrosion to the steel materials that are being used to construct the fuel cell.

— The fuel cell is popular for the generation of non-polluting products and its high-energy efficiency. The main drawbacks arise from the use of expensive noble metals such as platinum and palladium as electrodes. In addition, another factor for concern is the maintenance of the high-temperature condition in order to increase the electrical mobility of the sodium hydroxide electrolyte.

(b) Explain why this reaction is exothermic in terms of bond-breaking and bond-forming. Hence, describe how the strength of the bonds in the reactants is related to the strength of the bonds in the products for the above reaction.

Explanation:

$$2H-H + O=O \rightarrow 2H-O-H.$$

In the bond-breaking process, two H–H and one O=O bonds are broken, while four O–H bonds are formed during the bond-forming process. As the bond-forming process releases more energy than the bond-breaking process, the above reaction is exothermic.

An exothermic reaction would mean that the bonds that are formed in the products during the reaction are much stronger than the bonds that are broken in the reactants.

(c) Would you expect the heat change in the following reaction to be more exothermic, less exothermic, or the same as compared to the energy change in the first reaction? Explain your answer.

$$2H_2(g) + O_2(g) \rightarrow 2H_2O(g).$$

Explanation:

The heat change for the reaction, $2H_2(g) + O_2(g) \rightarrow 2H_2O(g)$, is less exothermic than that for the reaction, $2H_2(g) + O_2(g) \rightarrow 2H_2O(l)$. This is because in the latter reaction, the product is liquid water. When gaseous water condenses to liquid water, heat energy is evolved. This additional heat energy would add onto the heat change for the reaction, $2H_2(g) + O_2(g) \rightarrow 2H_2O(g)$, making the overall heat change less exothermic.

Do you know?

— State symbols are very important in a chemical equation. Different state symbols can result in different amounts of energy change for the reaction.

4. Soda water contains carbon dioxide dissolved in water, under pressure.

   (a) When a can of soda water is opened, the carbon dioxide gas escapes. Explain why the drink eventually goes "flat."

Explanation:

The fact that soda water contains carbon dioxide dissolved in water, under pressure, is an indication that the solubility of carbon dioxide in water is relatively low. The drink eventually goes "flat" because most of the carbon dioxide that were "forced" to dissolve would have escaped. The concentration of carbon dioxide that has dissolved in "flat" soda water is similar to that of plain water.

 Q   When carbon dioxide dissolves in water, how does it interact with the water molecules?

A: Although carbon dioxide is a non-polar molecule, the lone pair of electrons on both the oxygen atoms of CO_2 can form hydrogen bonds with the water molecules.

$$\delta- \longleftarrow + \quad \delta+ \quad + \longrightarrow \delta-$$
$$O = C = O$$

(b) If the amount of carbon dioxide dissolved decreases as temperature is increased, explain whether the dissolution of carbon dioxide is an exothermic or endothermic process.

Explanation:

The dissolution of carbon dioxide is an exothermic process. If the solubility is an endothermic process, then an increase in temperature should "encourage" solubility to take place as more energy is supplied for the dissolution process to absorb.

5. Fossil fuels are an important energy source but are non-renewable.

   (a) List the possible fossil fuels and discuss how they are used for electrical energy generation.

Explanation:

Fossil fuels include natural gas, coal, and oil. Different types of fuels, when burned, would give off different products:

Coal is mainly made up of carbon – $C(s) + O_2(g) \rightarrow CO_2(g)$ + heat.

Hydrocarbons such as methane (CH_4) – $CH_4(g) + 2O_2(g) \rightarrow CO_2(g) + 2H_2O(l)$ + heat.

The burning of fuel releases energy to boil water, and the steam that is produced is used to turn turbines which in turn generates electricity.

Is the energy that is used to boil the water fully extracted during the turning of turbines?

A: No. When steam cools down to form water, there is still heat energy in the water, which cannot be extracted anymore. So, when this water is released into rivers or streams, they can cause thermal pollution which affects aquatic life! Unlike the power plant, the fuel cell directly converts the chemical energy into electrical energy, hence there is less heat lost to the surroundings.

After the burning of coal, there is ash left over. What is ash really about?

A: A lot of students presume that ash is un-burned carbon; this is incorrect. Ash is mainly inorganic salt left behind after combustion. Fossil fuels are remains of living organisms which contain inorganic compounds other than organic compounds. As inorganic salts have a very high melting point, they are not easily vaporized.

(b) Discuss the advantages and disadvantages of using fossil fuels as compared to other renewable energy sources.

Explanation:

The disadvantage of using fossil fuels is due to environmental issues. Rampant burning of fossil fuels releases large amounts of carbon dioxide, of which the diminishing forests cannot re-absorb for photosynthesis. This causes global warming.

Another area of concern is that incomplete combustion of fossil fuels leading to the formation of carbon particles and carbon monoxide which are health hazards. In addition, fossil fuels are non-renewable sources and once depleted, there would be an energy crisis.

The main advantages of fossil fuels are: they are cheap, are easily obtainable, produce large amounts of energy, and are stable enough to be readily stored. In addition, the current infrastructure, plants and factories, and engines are mostly built to derive energy through the burning of fossil fuels. It is both economically and politically not viable to switch to other renewable energy sources.

(c) Suggest possible ways to solve the problems created by the use of fossil fuels.

Explanation:

We can plant more trees or conserve more forested areas to take in the additional amounts of carbon dioxide that are produced through the burning of fossil fuels.

Pollutants that are emitted from the combustion of fuel in car engines make up a substantial amount of man-made pollutants. Thus, to minimize emissions, cars are fitted with catalytic converters to remove three main pollutants (CO, NO_x, and non-combustible hydrocarbons) from exhaust gases. These pollutants are converted into less harmful products, such as CO_2, N_2, and H_2O, through a "three-way" catalytic converter.

To remove the emission of sulfur dioxide from power plants that use fossil fuels, we can employ the flue-gas desulfurization technique. A wet "scrubber" is used to extract the sulfur dioxide gas (or to scrub off) from the exhaust before it is being vented into the atmosphere. The chemical in the "scrubber" is a base while sulfur dioxide gas is acidic in nature. So, overall it is just an acid-base reaction.

The best ways would be to switch to other renewable energy sources such as hydroelectricity, solar energy, wind power, tidal power, geothermal heat, biomass fuel, and the fuel cell.

CHAPTER 6

RATE OF CHEMICAL REACTIONS

Do you know?

— Based on the Collision Theory, the particles need to "touch" each other in order to react. Although there are many collisions between the various particles, only a small fraction of them can result in a reaction. These are called *effective collisions*. The rate of reaction would therefore be expected to depend on the *frequency* of effective collisions between reactant particles.

Reactant molecules collide with correct geometry for fruitful reaction

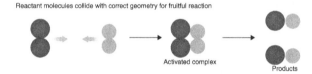

Activated complex

Products

Wrong collision geometry leads to unfruitful reaction

Molecules just bounce apart

— A chemical reaction involves the *rearrangement* of particles (atoms, ions, or molecules). In this rearrangement process, old bonds break and new bonds form. According to the Collision Theory, a number of conditions must be satisfied before a reaction can occur:

  (i) the reactant particles must *collide* with one another,
  (ii) with a *minimum amount of energy* (known as activation energy, E_a), and
  (iii) in the *correct orientation*.

(Continued)

(Continued)

As discussed in Chapter 1 on the Particulate Nature of Matter, we know that particles look like a cloud of electrons. During particulate collision, energy is needed to *overcome* the inter-electronic repulsion between the electron clouds of different particles. This required energy comes from the kinetic energy that the particles have. If the colliding particles do not have this *minimum* amount of energy, the collision will not lead to a reaction, i.e., the particles merely rebound from each other, like billiard balls bouncing away after collision. In addition, since there is a rearrangement of particles during the reaction, "old" bonds have to be broken. The energy that is required for the rearrangement of particles is part of the *activation energy* needed!

 Q Why is E_a labeled as the minimum energy needed? Shouldn't all the reactant particles have the same energy?

A: The particles in a system are constantly moving around, vibrating, or rotating in space. Energy is constantly being transferred from one particle to another when they collide with each other. As a result, not all particles would possess the same specific amount of energy at a specific time. But at a particular fixed temperature, there is always a *constant* distribution of kinetic energies. Well, you can understand the distribution of energy using the Maxwell–Boltzmann energy distribution curve that is shown below:

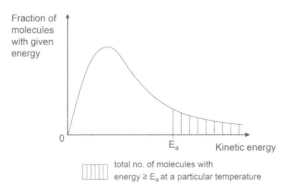

1. A piece of copper metal was introduced into a test tube of dilute nitric acid at 30°C. A slow reaction occurs.

   (a) If you wish to repeat the experiment and to increase the rate, what would be your choice?

      (I) Doubling the concentration of the nitric acid.
      (II) Doubling the volume of the nitric acid.
      (III) Doubling the temperature to 60°C.

Explanation:

Doubling the concentration of the nitric acid ensures more nitric acid per unit volume; this would increase the frequency of collisions between the nitric acid and the copper metal. Hence, the rate of reaction would increase.

Doubling the volume of the nitric acid does not increase the amount of nitric acid per unit volume; hence, it would not increase the rate of reaction.

Doubling the temperature to 60°C increases the kinetic energy of the particles. This ensures that more particles would have more kinetic energy to surmount the activation barrier. In addition, with a greater amount of kinetic energy, the frequency of collisions between the nitric acid and the copper metal would increase. These two factors would help to increase the rate of reaction.

Do you know?

— When the concentration of a reactant increases, the *frequency* of collisions will increase. When the reactant particles, especially those with kinetic energy that are greater than the activation energy (E_a),

(Continued)

(Continued)

bump into one another more often, the chances for a successful reaction to occur will be much higher, leading to an increase in the reaction rate.

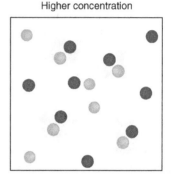

Low concentration   Higher concentration

— The average kinetic energy of the system is *directly proportional* to the temperature of the system.

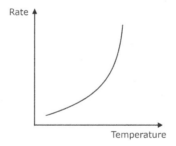

As temperature rises, the average kinetic energy of the reacting particles *increases*. This leads to an increase in the *frequency of collision*. In addition, the increase in temperature also leads to a significant increase in the number of reactant particles having an amount of kinetic energy greater than or equal to the E_a. Consequently, the *frequency of effective collisions* increases, and so, the reaction rate increases. The evidence of

(Continued)

(Continued)

this is in the Maxwell–Boltzmann distribution curve that is shown below:

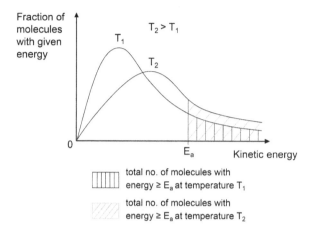

In general, *increasing the temperature increases the rate of a reaction.* For many reactions, the rate is *approximately doubled* for every 10°C rise in temperature.

(b) Which of your choice will probably increase the rate the most?

Explanation:

Doubling the temperature to 60°C would increase the rate the most. This is because if we use the guideline that for many reactions, the rate is approximately doubled for every 10°C rise in temperature, then an increase of the temperature by 30°C would lead to an eight-fold ($2 \times 2 \times 2$) increase in the rate of reaction.

(c) Which conidition is likely to increase the rate as well as increase the amount of product formed?

Explanation:

Doubling the temperature to 60°C is likely to increase the rate as well as increase the amount of product formed. This is because at a higher temperature, more particles have kinetic energies to overcome the activation barrier, hence leading to the formation of more products.

> (d) If the volume of the piece of copper metal is doubled, how does this affect the rate of the reaction? Explain.

Explanation:

If the volume of the piece of copper metal is doubled, the surface area would also be doubled. The rate of the reaction would increase as the nitric acid has a greater frequency of collisions with the increase in surface area.

 Will the increase in the rate of reaction be significant?

A: Maybe not. Relatively, the increase in the rate of reaction by doubling the volume would not be as significant as if the piece of copper metal had been cut up into many smaller pieces.

Do you know?

— Reactions involving liquids or gases occur much faster than those involving solids. Dissolving a substance is a *means* of providing a much greater surface area for reactions to take place, as we are dealing with the smallest particles present — the ions or molecules.

(1) $NaCl(s) + AgNO_3(s) \rightarrow$ *no reaction.*

(2) $NaCl(aq) + AgNO_3(aq) \rightarrow NaNO_3(aq) + AgCl(s)$ *immediate reaction.*

The ratio of surface area to mass is greater in small particles than in large particles. This would imply that the surface area over which the solid can come in contact with liquid or gaseous reactants is greater.

(e) If a sugar cube is held in the flame of a candle, the sugar melts and turns brown but does not burn. However, the cube will burn if copper powder is first rubbed into it, even though the copper does not react. Explain how the copper helps in the combustion of sugar in oxygen.

Explanation:

Copper metal is a good conductor of heat. As a result, heat energy is transferred much more rapidly to the sugar. In addition, the mixing of the copper powder with the sugar ensures that heat energy is transferred to all the sugar at the same time. With all the sugar being subjected to a high temperature at the same time, the sugar is likely to burn.

2. A small amount of manganese(IV) oxide powder was added to $100\,cm^3$ of $0.4\,mol\,dm^{-3}$ hydrogen peroxide solution at 30°C. The equation for the reaction is as follows:

$$2H_2O_2(aq) \rightarrow O_2(g) + 2H_2O(l).$$

The volume of oxygen gas produced was measured at regular time intervals. The result is shown in Graph 1.

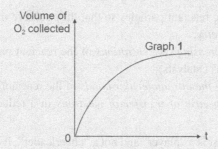

(a) Explain the role of manganese(IV) oxide.

Explanation:

The manganese(IV) oxide was acting as a catalyst, which speeds up the rate of the reaction. At the end of the reaction, the catalyst remained unchanged.

Do you know?

— A catalyzed reaction proceeds via an *alternative pathway* of lower E_a in contrast to the uncatalyzed reaction. The energy profile diagrams for a catalyzed and uncatalyzed reactions are shown below:

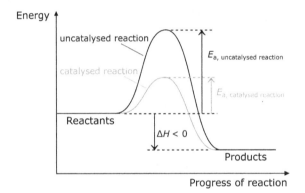

— A catalyst does its job by:
  - *orientating* reactant particles so that they achieve the *correct collision geometry*;
  - *locally increasing concentrations* of the reactant particles (for heterogeneous catalysts);
  - *weakening the intramolecular bonds* of the reactant molecules; and
  - *facilitating ease of transfer of electrons* in a reduction–oxidation reaction.

Thus, a catalyst is a "player" and not a "cheerleader"; it does participate in the reaction. The pathway whereby the catalyst is acting as a "reactant" is *different* from the one without the catalyst's participation. Hence, these two pathways have *different* activation energies.

How would you use a Maxwell–Boltzmann distribution curve to explain how a catalyst speeds up the rate of a reaction?

A: The Maxwell–Boltzmann distribution curve below shows the distribution of kinetic energies of reacting particles and the activation energies of both the catalyzed and uncatalyzed reactions.

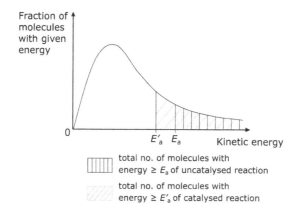

Have you noticed that now we have a greater proportion of reacting particles having sufficient kinetic energy greater than the *lowered* E'_a, for the catalyzed reaction? This is indicated by the larger shaded area in the diagram.

Thus, having a greater number of reactant particles with kinetic energy greater than or equal to the lowered activation energy (E'_a), results in an increase in the *frequency of effective collisions*. This leads to an increase in the reaction rate.

 What is a heterogeneous catalyst?

A: When the catalyst and the reactants are not of the same phase, the catalyst is termed a "heterogeneous catalyst," and the reaction is termed "heterogeneous catalysis." So, a homogeneous catalyst is of the same phase as the reactants. The following schematic shows the role of the heterogeneous catalyst, nickel, during the hydrogenation of alkene:

1. Adsorption of reactants

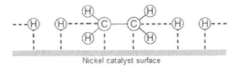

Nickel catalyst surface

2. Chemical reaction

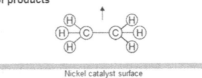

Nickel catalyst surface

3. Desorption of products

Nickel catalyst surface

(b) With the aid of an experiment, explain how you would verify your answer for part (a).

Explanation:

In another experiment, the volume of oxygen gas produced was measured at regular time intervals for an uncatalyzed $100 \, cm^3$ of $0.4 \, mol \, dm^{-3}$ hydrogen peroxide solution at 30°C. The graph for the uncatalyzed hydrogen peroxide solution should have a less steep gradient of change as compared to the catalyzed reaction as shown below:

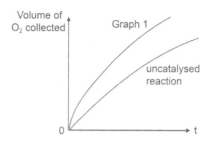

Q Would the two graphs show the same volume of O_2 collected at the end of the two experiments?

A: Certainly, this is because you have used the same volume and same concentration of hydrogen peroxide solution.

Do you know?

— The rate of reaction is defined as the *change in the amount of a reactant or product with time*. Mathematically, it can be expressed as:

$$\text{Rate of reaction} = \frac{\text{amount of product formed}}{\text{time taken}}$$

$$\text{or Rate of reaction} = -\frac{\text{amount of reactant used up}}{\text{time taken}}$$

The negative sign is to account for the decreasing amount of reactant over time.

— Experimentally, if during the course of the reaction, the mass of the system changes, then the rate of reaction can be monitored by measuring the mass of the system with respect to the passage of time. An example is the reaction between hydrochloric acid and marble chips ($CaCO_3$). As one of the products is carbon dioxide gas, the total mass

(Continued)

(*Continued*)

of the system decreases when the carbon dioxide escapes. The amount of the carbon dioxide that is escaping is proportional to the amount of reactant decreasing or inversely proportional to the amount of unreacted reactants.

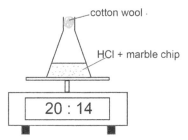

Since the product is a gas, then the rate of reaction can also be determined by collecting the gas that is produced with respect to time:

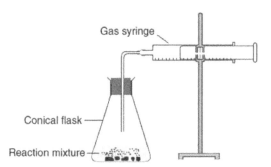

When experimental values of the amount and time are plotted graphically, we would obtain the following graphs:

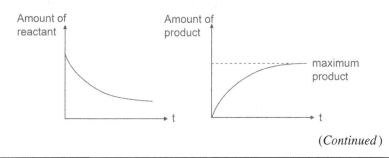

(*Continued*)

(Continued)

— From the amount–time graph, the rate of reaction is obtained by calculating the gradient of the tangent drawn on the curve at $t = 0$.

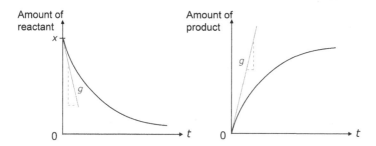

The steeper the gradient of the graph at any particular time, the faster the rate of reaction at that time. Thus, the two graphs show that the reaction rate is fastest at the start of the reaction when the reactants are at their highest concentration. As time progresses, the *amount of reactant decreases,* the *gradient* becomes less steep (or less negative) *with time*, hence the *rate of reaction decreases with time*. When the gradient is zero, the reaction has stopped.

Q   Can we calculate the rate at any point on the graph? Or is it necessary to only determine it at $t = 0$?

A: Well, if you only want to determine the rate of reaction at any point on the graph, then you can simply draw the tangent at the point of interest and then calculate the gradient of the tangent. But if the reaction equation is not a simple one, then you can only use the rate at $t = 0$. For more details, you can refer to *Understanding Advanced Physical Inorganic Chemistry* by J. Tan and K.S. Chan.

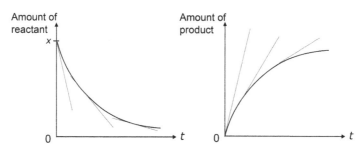

(c) Analyze the graph and explain why the shape of the graph is as shown.

Explanation:

Initially, the amount of hydrogen peroxide is at its maximum, hence the rate of the amount of O_2 evolved is also at its maximum. As time passes, the amount of hydrogen peroxide decreases. Thus, the rate of the amount of O_2 that is evolved decreases. The maximum amount of O_2 at the end of the experiment is half of the initial amount of hydrogen peroxide present since two moles of H_2O_2 gives one mole of O_2.

 How would you interpret the shape of the graph if it is a straight linear line?

A: This is not a difficult task to do. Do you know that the gradient of the graph at each point represents the rate of the reaction, and at each point of the graph, there is also a specific concentration value for the reactant? As the reaction proceeds, the concentration of the reactant decreases, and if you have a straight linear line, this would mean that the rate of the reaction is the same ALTHOUGH the concentration of the reactant is decreasing. The implication is that the rate of the reaction is INDEPENDENT of the concentration of the reactant!

 So, if we have a curved plot, does that mean that as the reaction proceeds, the concentration of the reactant decreases and because the rate of the reaction is dependent on the concentration of the reactant, the rate also changes with time?

A: Absolutely spot on! Yes, the dependency of the rate on the concentration of the reactant is reflected as a trend of changing gradients, hence you have a curved plot.

(d) If it takes 10 min to collect $75\,cm^3$ of oxygen, calculate the average speed of the reaction in terms of the volume of oxygen collected.

Explanation:

Average speed of the reaction $= \dfrac{75}{10} = 7.5\,\text{cm}^3\,\text{min}^{-1}$.

 Q So, $7.5\,\text{cm}^3\,\text{min}^{-1}$ is not the speed of the reaction at each point of the graph?

A: Of course not. This is because the graph is a curve and not a straight line.

(e) On the same axes above, sketch the graphs that you would expect if the experiment is repeated, with the following changes in each case:

(i) Using $50\,\text{cm}^3$ of $0.8\,\text{mol}\,\text{dm}^{-3}$ hydrogen peroxide.
(ii) Using $200\,\text{cm}^3$ of $0.4\,\text{mol}\,\text{dm}^{-3}$ hydrogen peroxide.
(iii) Using half the amount of manganese(IV) oxide.
(iv) Heating the mixture to 40°C.

Explanation:

(i) The concentration of the hydrogen peroxide ($0.8\,\text{mol}\,\text{dm}^{-3}$) is two times more than that for the initial experiment. Hence, the rate of the reaction should be doubled. But since the volume ($50\,\text{cm}^3$) is half of that for the initial experiment, the volume of oxygen that is collected at the end of the reaction is the same as that for the initial experiment.

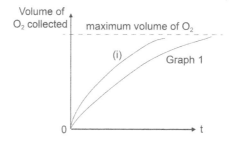

(ii) The concentration of the hydrogen peroxide ($0.4\,mol\,dm^{-3}$) is the same as that for the initial experiment. Hence, the rate is the same. But since the volume ($100\,cm^3$) is two times more than that for the initial experiment, the volume of oxygen that is collected at the end of the reaction is doubled that for the initial experiment.

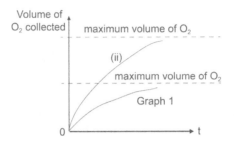

(iii) Using a smaller amount of catalyst would decrease the rate of the reaction, but the volume of oxygen that is collected at the end of the experiment is the same.

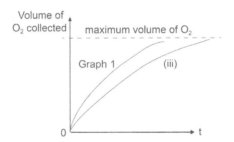

(iv) Heating the mixture to 40°C would approximately double the rate, but the volume of oxygen that is collected at the end of the experiment is the same as that for the initial experiment.

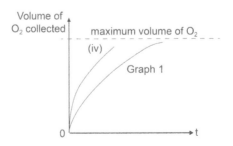

3. A piece of zinc metal of 5 cm in length was each added to five beakers containing the following acid solutions:

Beaker 1: 50 cm^3 of 1.0 mol dm^{-3} hydrochloric acid.
Beaker 2: 100 cm^3 of 1.0 mol dm^{-3} hydrochloric acid.
Beaker 3: 50 cm^3 of 1.0 mol dm^{-3} sulfuric acid.
Beaker 4: 25 cm^3 of 1.0 mol dm^{-3} sulfuric acid.
Beaker 5: 50 cm^3 of 1 mol dm^{-3} ethanoic acid.

In each case, the zinc metal reacted completely.

(a) Give the ionic equation between zinc and acid.

Explanation:

$$Zn(s) + 2H^+(aq) \rightarrow Zn^{2+}(aq) + H_2(g).$$

Do you know?

— When matter reacts, not all the particles (atoms, ions, or molecules) participate in the reaction. An *ionic equation* is a chemical equation which shows specifically which atoms, ions, or molecules take part in the reaction. You can say that an ionic equation is an "in-a-nutshell" equation!

— When writing an ionic equation from a given chemical equation, the purpose is to get rid of the *spectator* ions. As these spectators ions do not participate in the chemical reaction, they are present just to maintain electrical *neutrality* for the system.

Q So, how was the above ionic equation derived?

A: Step 1: Write the balanced chemical equation.

$$Zn + 2HCl \rightarrow ZnCl_2 + H_2.$$

Step 2: Put in the appropriate state symbols.

$$Zn(s) + 2HCl(aq) \rightarrow ZnCl_2(aq) + H_2(g).$$

Step 3: Show all the free ions in the aqueous solution.

$$Zn(s) + 2H^+(aq) + 2Cl^-(aq) \rightarrow Zn^{2+}(aq) + 2Cl^-(aq) + H_2(g).$$

Step 4: Eliminate the spectator ions on both sides of the equation.

$$Zn(s) + 2H^+(aq) + 2Cl^-(aq) \rightarrow Zn^{2+}(aq) + 2Cl^-(aq) + H_2(g).$$

Step 5: You have the ionic equation!

$$Zn(s) + 2H^+(aq) \rightarrow Zn^{2+}(aq) + H_2(g).$$

(b) Sketch the graph of volume of gas collected against time for each of the five reactions.

Explanation:

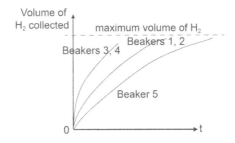

(c) Explain why the rate of reaction is different in each beaker.

Explanation:

Beaker 1 ($50\,cm^3$ of $1.0\,mol\,dm^3$ hydrochloric acid) and Beaker 2 ($100\,cm^3$ of $1.0\,mol\,dm^{-3}$ hydrochloric acid) have the same graph because the concentration of the two acid solutions are the same. The amount of acid in Beaker 1 is more than sufficient to react with the zinc metal; thus, the volume of the acid is not an issue here.

Beaker 3 ($50\,cm^3$ of $1.0\,mol\,dm^{-3}$ sulfuric acid) and Beaker 4 ($25\,cm^3$ of $1.0\,mol\,dm^{-3}$ sulfuric acid) have the same graph for the same reason as that for Beakers 1 and 2. But the graph for Beakers 3 and 4 show a greater rate of reaction when compared with that for Beakers 1 and 2. This is because although both concentrations of hydrochloric acid and sulfuric acid are the same, the concentration of H^+ ions in sulfuric acid is twice of that in hydrochloric acid. Hence, the rate of reaction is faster in Beakers 3 and 4 than in Beakers 1 and 2.

As for Beaker 5 ($50\,cm^3$ of $1\,mol\,dm^{-3}$ ethanoic acid), the ethanoic acid is a weak acid compared to the hydrochloric acid. As weak acid molecules do not fully dissociate in water, the amount of H^+ ions is lower than that in a strong acid solution. Hence, the rate of reaction in Beaker 5 is slower.

(d) How would you expect the rate of reaction to be like if the piece of zinc metal is cut into smaller pieces for Beaker 1?

Explanation:

The rate of reaction should increase as the surface area for the reaction to take place has increased. The increase in surface area increases the frequency of collisions between the acid and the metal.

(e) If $100\,cm^3$ of gas was collected when the zinc completely reacted in Beaker 1, what would be the volume of gas collected for the other four beakers?

Explanation:

All five beakers would give the same volume of hydrogen gas at the end of the experiments as the same amount of zinc metal is used, and the metal is the limiting reagent.

(f) Calculate the mass of magnesium needed in order to collect $100\,cm^3$ of gas.

Explanation:

Assuming it is at r.t.p. conditions, 1 mole of gas occupies a volume of $24.0 \, dm^3$. $100 \, cm^3 = 0.1 \, dm^3$.

Amount of H_2 in $100 \, cm^3 = \dfrac{0.1}{24.0} = 4.17 \times 10^{-3} \, mol$.

One mole of magnesium gives one mole of H_2; the amount of magnesium used is $4.17 \times 10^{-3} \, mol$.

Mass of magnesium used $= 4.17 \times 10^{-3} \times 24.3 = 0.101 \, g$.

---

4. When aqueous sodium thiosulfate, $Na_2S_2O_3$, reacts with sulfuric(VI) acid as shown in the set-up below, sulfur is precipitated:

$$Na_2S_2O_3(aq) + H_2SO_4(aq) \rightarrow S(s) + Na_2SO_4(aq) + SO_2(g) + H_2O(l).$$

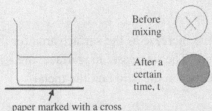

Looking down into the beaker:

Before mixing

After a certain time, t

paper marked with a cross

$20 \, cm^3$ of $0.10 \, mol \, dm^{-3}$ sodium thiosulfate was poured into a beaker and $20 \, cm^3$ of $0.15 \, mol \, dm^{-3}$ sulfuric(VI) acid was then added. The temperature of the experiment was maintained at 30°C. The time taken for the sulfur to cover the cross was recorded. This is called Experiment 1.

The experiment was repeated four times using the same volumes of $0.10 \, mol \, dm^{-3}$ sodium thiosulfate but different concentrations of sulfuric(VI) acid. In each experiment, the temperature and the concentration of the acid were varied. The table below shows the results of the experiments.

(Continued)

	(Continued)		
Experiment	Concentration of acid/(mol dm^{-3})	Temperature (°C)	Time taken (s)
1	0.15	30	60
2	0.10	40	40
3	0.10	30	80
4	0.05	40	50
5	0.05	30	100

(a) Select two experiments you would use in order to show the effect of acid on the speed of reaction.
(b) Explain your choice in part (a).

Explanation:

In order to simply show the effect of acid on the speed of reaction, the temperature must be constant for both the experiments. This is to ensure that there is only one independent variable that has been changed. We should select Experiments 1 and 3, 1 and 5, or 3 and 5.

(c) What conclusions can you derive from the results of all these experiments?

Explanation:

The longer the time taken for the "cross" mark to be masked, the slower the rate of reaction, i.e., rate $\propto \frac{1}{\text{time taken}}$. If we compare Experiments 1 and 5, when the concentration of the acid is decreased three-fold, the time taken increases from 60 s to 100 s. This shows that a lower concentration of the acid gives a lower rate of reaction.

(d) How would you explain your conclusions for part (c) based on the Collision Theory?

Explanation:

A lower concentration of acid has a smaller amount of H^+ ions, hence the frequency of collisions between the reactants decreases. This thus lowers the rate of reaction.

(e) Suggest another experimental set that you would use in order to evaluate the effect of the concentration of sodium thiosulfate on the speed of reaction.

Explanation:

To evaluate the effect of the concentration of sodium thiosulfate on the speed of reaction, we need to keep both the temperature and concentration of sulfuric(VI) acid the same for both experiments. Thus, for the other experimental set, we would use:

$$20 \, cm^3 \text{ of } 0.05 \, mol \, dm^{-3} \text{ sodium thiosulfate with } 20 \, cm^3$$
$$\text{of } 0.15 \, mol \, dm^{-3} \text{ sulfuric(VI) acid at } 30°C.$$

(f) Give the ionic equation for the above reaction.

Explanation:

$$S_2O_3^{2-}(aq) + 2H^+(aq) \rightarrow S(s) + SO_2(g) + H_2O(l).$$

(g) Draw the dot-and-cross diagrams for both SO_2 and H_2O.

Explanation:

$$\ddot{O}_{xx}^{xx}\ddot{S}:\ddot{O}_{xx}^{xx} \quad H:\ddot{O}:H$$

5. Limestone contains calcium carbonate. Excess limestone was reacted with dilute hydrochloric acid:

$$CaCO_3(s) + 2HCl(aq) \rightarrow CaCl_2(aq) + H_2O(l) + CO_2(g).$$

The rate of reaction was followed by measuring the mass lost during the reaction using a set-up as shown on page 116. The results were tabulated below:

Time t/min	Total mass lost/g
0	0.00
4	0.18
8	0.30
12	0.38
16	0.44
20	0.48
24	0.51

(a) Sketch a graph of total mass lost against time.

Explanation:

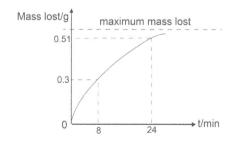

(b) Explain the shape of the graph in part (a).

Explanation:

Initially, the mass of calcium carbonate is at its maximum; the mass lost is zero at $t = 0$. The rate of the mass lost is also at its maximum. As time passes, the mass of calcium carbonate decreased. Thus, the rate of the mass lost decreases. The maximum mass lost at the end of the experiment is directly proportional to the initial mass of calcium carbonate used.

(c) If 0.6 g of calcium carbonate was used initially, sketch the graph of the mass of calcium carbonate against time.

Explanation:

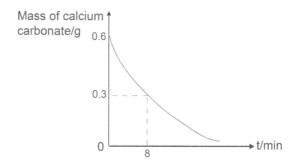

(d) Explain the relationship between the two graphs in parts (a) and (c).

Explanation:

The graphs in parts (a) and (c) are mirror images of each other. This is logical as the gradients of the graph in part (a) show the rate of mass lost, which is how fast the calcium carbonate had reacted in part (c).

(e) Using the Collision Theory, explain why the rate of the reaction decreases.

Explanation:

As the mass of the calcium carbonate decreases, there is a decreasing amount of the calcium carbonate present. Hence, the frequency of collisions between the acid and the carbonate decreases.

(f) Draw another labeled diagram to show a different method of following the rate of reaction of limestone and hydrochloric acid.

Explanation:

The rate of reaction of limestone and hydrochloric acid can be observed by collecting the carbon dioxide produced, measuring the volume of the carbon dioxide gas with respect to time as follows:

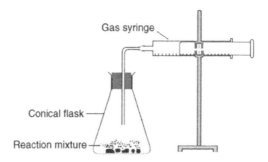

(g) Give the ionic equation for the reaction between calcium carbonate and hydrochloric acid. Hence, identify the spectator ions.

Explanation:

$$CaCO_3(s) + 2H^+(aq) \rightarrow Ca^{2+}(aq) + H_2O(l) + CO_2(g).$$

The spectator ions are Cl^- and Ca^{2+}.

---

6. Coal mines are hazardous places because of the danger of fires from the coal dust, and also from the possible presence of methane gas (CH_4).

   (a) Explain why fine particles of coal dust are more reactive as compared to lumps of coal.

Explanation:

Fine particles of coal dust have a greater surface area for reaction as compared to lumps of coal. Hence, the frequency of collisions between the coal dust and oxygen is higher. This leads to greater rate of reaction.

   (b) Give a chemical equation for the reaction between coal dust and excess oxygen.

Explanation:

$$C(s) + O_2(g) \rightarrow CO_2(g).$$

Q Why is the coal dust not converted to $CO(g)$?

A: The question stated "excess oxygen," so we need to assume that it is a complete combustion.

   (c) List the likely products that would be formed if oxygen is in limited supply.

Explanation:

Carbon monoxide would be formed if oxygen is in limited supply.

> (d) Is this reaction in part (b) exothermic or endothermic? Explain your answer.

Explanation:

The reaction in part (b) is exothermic as the carbon dioxide formed is more stable than the reactants.

> (e) Write a chemical equation for methane combusting in excess oxygen.

Explanation:

$$CH_4(g) + 2O_2(g) \rightarrow CO_2(g) + 2H_2O(l).$$

> (f) Given the following bond energies:
>
> | Covalent bond | Bond energy in kJ mol^{-1} |
> |:---:|:---:|
> | H–C | 410 |
> | C–C | 350 |
> | O=O | 610 |
> | O–H | 460 |
> | C=O | 740 |
>
> Calculate the ΔH for the combustion of methane in part (e), assuming that the physical state of water is in the gaseous form.

Explanation:

$$H-\underset{\underset{H}{|}}{\overset{\overset{H}{|}}{C}}-H \quad + \quad 2O=O \quad \longrightarrow \quad O=C=O \quad + \quad 2H-O-H$$

Total energy absorbed during bond breaking

$$= 4 \times BE(C–H) + 2 \times BE(O=O)$$
$$= 4 \times 410 + 2 \times 610$$
$$= 2860 \text{ kJ mol}^{-1}.$$

Total energy released during bond forming

$$= 2 \times BE(C=O) + 4 \times BE(O–H)$$
$$= 2 \times 740 + 4 \times 460$$
$$= 3320 \text{ kJ mol}^{-1}.$$

ΔH = (Sum of total energy for bond breaking)
– (Sum of total energy for bond making).

Hence, $\Delta H = 2860 - 3320 = -460 \text{ kJ mol}^{-1}$.

CHAPTER 7

EQUILIBRIA, AMMONIA, AND SULFUR

Do you know?

— Not all chemical reactions would proceed to completion, and not all reactions are irreversible. In fact, there are many chemical reactions that are reversible.

— A *reversible* reaction can proceed in *two* directions: forward and backward. Equations of reversible reactions are represented with a double-headed arrow ($\rightleftharpoons$):

$$A + B \rightleftharpoons C + D.$$

The reaction from the left to the right is known as the *forward* reaction.

The reaction from the right to the left is known as the *backward* reaction or the *reverse* reaction.

— As reversible reactions tend not to go to completion, a mixture of both the reactants and the products is obtained no matter how long you let the reaction carry on. When the concentration of the reactants and products are *constant*, we say that the mixture is at *equilibrium*.

Consider the following reaction: $A + B \rightleftharpoons C + D$.

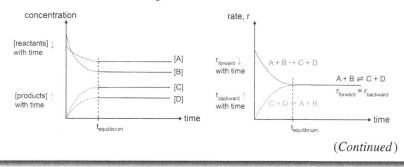

(Continued)

133

> (*Continued*)
>
> At the point of equilibrium, at $t_{equilibrium}$, both the forward and backward reactions occur at the same rate, i.e., forward rate (R_f) = backward rate (R_b).
>
> More specifically, the system is said to be in *dynamic equilibrium*. As the rates of these reactions are equal, it may seem that the reactions have completely stopped but this is not true. Although the concentrations of the substances remain unchanged (as indicated by the term "equilibrium"), there is still *activity* going on; both forward and backward reactions are continually occurring (as indicated by the term "dynamic"), but since they proceed at the same rate, each species is formed as fast as it is consumed, resulting in a constant concentration term.

Q   What would likely make a reaction to be a reversible one?

A: The reversibility of a reaction depends on the *magnitude* of the activation energy (E_a) of the backward reaction. For an *exothermic* reversible reaction, if the E_a of the backward reaction is too large, the backward reaction is essentially non-occurring as compared to the forward reaction. Hence, one can assume that the forward reaction would proceed almost to completion. As for an *endothermic* reversible reaction, the higher E_a for the forward reaction would make it less likely for it to proceed as compared to the backward reaction.

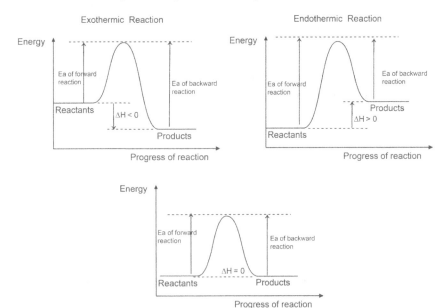

The energy profile diagram for a reversible reaction shows that the E_a of the backward reaction is *comparable* to that of the forward reaction. Or in short, the ΔH of reaction is about zero!

 Q Can equilibrium be achieved if the reactant or product is allowed to escape?

A: No! Equilibrium can only occur in what we call a *closed* system, where particles are not allowed to move in or out of the system. Just like if you placed a piece of wet cloth in a closed container, the cloth would never dry up. The rate of evaporation will be equal to the rate of condensation. BUT if this piece of cloth is placed in a windy place, basically there is no condensation but only evaporation as the wind would collide with the vapor molecules and "sweep" them away.

 Q When equilibrium is reached, would the amount of reactants be the same as the amount of products?

A: Not necessarily. The amount of reactants and products at equilibrium really depends on the *nature* of the reaction PLUS the *ratio* of the concentrations of the reactants that you have started with. Refer to *Understanding Advanced Physical and Inorganic Chemistry* by J. Tan and K.S. Chan, if you are interested to know more.

 Q Are the following two reactions the same?

$$A + B \rightleftharpoons C + D.$$

$$C + D \rightleftharpoons A + B.$$

A: They are not the same. For the first reaction, the forward reaction is $A + B \rightarrow C + D$, whereas the backward reaction is $C + D \rightarrow A + B$. This is opposite of that for the second reaction, which has the forward reaction, $C + D \rightarrow A + B$.

1. Nitrogen is converted to nitrogen monoxide by oxygen from the air in car engines. The reaction is a reversible one:

$$N_2(g) + O_2(g) \rightleftharpoons 2NO(g), \quad \Delta H = +180 \text{ kJ mol}^{-1}.$$

In order to minimize the yield of nitrogen monoxide at equilibrium, it is necessary to control the conditions very carefully.

(a) Draw the energy profile diagram indicating the activation energies and ΔH on the diagram.

Explanation:

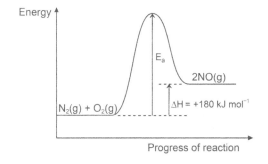

Q What is the difference between an energy level diagram and an energy profile diagram?

A: If we compare the energy level diagram in Chapter 5 on Energy Change and Fuels and the energy profile diagram here, you would notice that the x-axis of the energy level diagram is not labeled but that for the energy profile diagram is labeled as "Progress of reaction" or "Reaction coordinate." What is the meaning of this? The big difference between these two diagrams is that the energy profile diagram shows you the change in energy content of the system as the reaction progresses from the reactants to the products. Hence, the phrase "progress of reaction." In contrast, the energy level diagram simply gives you the energy content of the reactants and products, without informing you about the transformation of the energy content in the process of the reaction.

(b) Using the following information:

Covalent bond	Bond energy in kJ mol^{-1}
N≡N	994
O=O	496

Calculate the bond energy of the nitrogen–oxygen bond in nitrogen monoxide.

Explanation:

$$N \equiv N + O = O \rightarrow 2N = O.$$

Let the bond energy of the nitrogen–oxygen bond in nitrogen monoxide be BE(N=O).

Total energy absorbed during bond breaking $= BE(N \equiv N) + BE(O=O)$
$$= 994 + 496$$
$$= 1490 \text{ kJ mol}^{-1}.$$

Total energy released during bond-forming $= 2 \times BE(N=O)$.

$\Delta H =$ (Sum of total energy for bond breaking)
      $-$ (Sum of total energy for bond making).

Hence, $\Delta H = 1490 - 2 \times BE(N=O) = +180 \text{ kJ mol}^{-1}$
$$\Rightarrow 2 \times BE(N=O) = 1490 - 180$$
$$\Rightarrow BE(N=O) = 655 \text{ kJ mol}^{-1}.$$

(c) Explain why the above reaction is endothermic in nature.

Explanation:

The reaction is endothermic in nature as the nitrogen monoxide formed is less stable than the reactants.

Q The endothermic nature of the reactions indicates that it has absorbed energy. From this perspective, why is energy needed for the above reaction?

A: Easy! If you look closely at the bonds that need to be broken and the bonds that are formed, you would notice that you need to break a N≡N triple bond in exchange for a N=O double bond. Assuming that the O=O double bond is as strong as the N=O double bond, the energy needed to break the N≡N is not "compensated" fully by the energy that is released when the N=O double bond is formed. Hence, in total, the reaction still need extra energy in order to react. This accounts for the endothermic nature of the reaction.

(d) State, with a reason, whether a high or low temperature should be used.

Explanation:

As the reaction is endothermic in nature, according to Le Chatelier's Principle, a high temperature would shift the position of equilibrium toward the right. This would favor the formation of the product.

Do you know?

— Le Chatelier's Principle is a helpful rule to predict the effects of changes applied to a system that is *already* at equilibrium. It helps us to predict in which *direction* the equilibrium *position* will shift (favoring either the forward or backward reaction) in response to the change. It states that,

"When a system in *equilibrium* is subjected to a change, the system would *respond* in such a way so as to *counteract* the imposed change and *re-establish* the equilibrium state."

— When a change in conditions is introduced to an equilibrium system, the system will no longer be in equilibrium since the change will affect the rates of both the forward and backward reactions to different extents. The system will then readjust itself to attain a new equilibrium where the concentrations of reactants and products become constant again. These concentration values would, however, be *different from those of the previous equilibrium state*.

Q Why would the position of equilibrium shift right for an endothermic reaction when temperature is increased? Does this mean that the position of equilibrium would shift left for an exothermic reaction if the temperature is increased?

A: We can use kinetics to understand why the position of equilibrium would shift as such! When the temperature of the system is increased, the rates of both the forward and backward reactions would increase. But the percentage increase is greater for the reaction that has the greater activation energy, i.e., the forward reaction for the endothermic reaction that we are discussing here. So, as time passes, the position of equilibrium would shift *right*. This explains how the system "gets rid" of the added heat. (For mathematical proof, refer to *Understanding Advanced Physical Inorganic Chemistry* by J. Tan and K.S. Chan). Thus, for an exothermic reaction, since the backward reaction has a higher activation energy than the forward reaction, the position of equilibrium would shift to the left.

Q How do you use kinetics to explain how the system responds to a decrease in temperature for an exothermic reaction?

A: When the temperature of the system is decreased, the rates of both the forward and backward reactions would decrease. But the percentage decrease is greater for the reaction that has the greater activation energy, i.e., the backward reaction for the reaction here. So, as time passes, the position of equilibrium would shift *right*. This explains how the system "supplies" more heat.

Do you know?

— In summary:

an increase in temperature favors the endothermic reaction; and
a decrease in temperature favors the exothermic reaction.

(e) State, with a reason, what effect, if any, an increase in pressure would have on the equilibrium yield.

Explanation:

According to Le Chatelier's Principle, when pressure is increased for a system, the system would respond in such a way so as to reduce the pressure by producing a smaller number of particles. For the above reaction, as the numbers of particles on both sides of the equation are the same, the position of equilibrium would not shift. Thus, there is no change on the equilibrium yield.

 How would you use kinetics to explain how the system responds to the change?

A: When the total pressure increases, the volume of the system shrinks. The gaseous particles now occupy a smaller volume. As a result, the effective collisional frequencies of both the forward and backward reactions increase. These lead to an increase in the rates of both the forward and backward reactions. The percentage increase is the same for both directions of the reaction equation. So, as time passes, the position of equilibrium would not shift to any side.

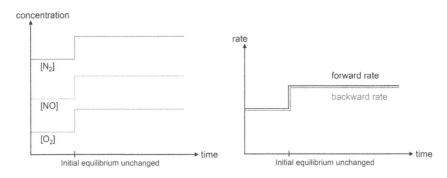

(f) Draw the dot-and-cross diagram for nitrogen monoxide.

Explanation:

$$:\overset{\bullet}{N}:\overset{\times\times}{\underset{}{O}}\overset{}{\underset{}{}}$$

> **Q** Why does the nitrogen atom of the nitrogen monoxide not follow the octet rule?

A: The octet rule is just a guideline! Not every element has to follow the octet rule when it reacts. There are many stable compounds that can exist although the element in the compound may not have followed the octet rule.

(g) Explain why nitrogen monoxide is a reactive species.

Explanation:

If we use the octet rule as a guideline, the nitrogen atom of nitrogen monoxide is an electron-deficient species as it does not have an octet configuration. Hence, it is a reactive species. In fact, nitrogen monoxide can undergo dimerization to form N_2O_2:

$$:\overset{\bullet\bullet}{O}:\overset{\times}{N}\overset{\times}{}$$
$$:\overset{\times}{N}:\overset{}{O}\overset{\times}{\underset{\times\times}{}}$$

> **Q** Wait a minute, is there a better way other than using the octet rule to explain why the nitrogen monoxide dimerizes?

A: Easy! When the nitrogen monoxide undergoes dimerization, an N–N bond is formed between the two molecules. What is the consequence of the formation of this bond? When a bond forms, energy is evolved! Hence, the energy level of the dimer is more stable than the nitrogen monoxide. This is the actual driving force for the reaction to happen and not the octet rule!

2 Sulfuric(VI) acid is manufactured by the Contact Process, using vanadium(V) oxide as a catalyst:

$$2SO_2(g) + O_2(g) \rightleftharpoons 2SO_3(g), \quad \Delta H < 0.$$

(a) Explain the meaning of the ΔH.

Explanation:

The enthalpy change of a reaction, ΔH, indicates the relative energy level of the reactants and products. An exothermic ΔH indicates that the products have a lower energy level than the reactants. Thus, the products are energetically more stable than the reactants. It is likewise for an endothermic ΔH.

Q   What role does the catalyst play in an equilibrium reaction?

A: A catalyst only aids a reversible reaction to attain equilibrium in a shorter amount of time. It does not cause any change to the position of the equilibrium, i.e., the use of a catalyst does not give you more product as compared to an uncatalyzed reaction. This is because a catalyst lowers the activation energies of both the forward and backward reactions to the same extent. This, in turn, would lead to the rates of both the forward and backward reactions to be increased to the same extent.

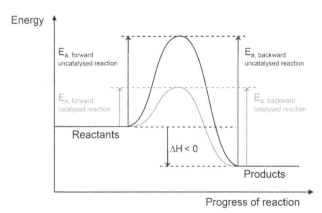

(b) What does the sign tell you about the strength of the bonds within the reactants and products?

Explanation:

For an exothermic ΔH, the products are more stable than the reactants. Actually, this is because the strengths of the bonds that are within the products are stronger than those in the reactants.

(c) State two other conditions that are necessary for a good yield of sulfur trioxide (SO_3).

Explanation:

The conditions that are necessary for a good yield of sulfur trioxide (SO_3) are:

- Temperature: 400°C
  Reason for choice: Lowering the temperature of the reaction vessel will favor the forward reaction since it is exothermic, and hence, a higher yield of $SO_3(g)$ can be obtained.

  On the other hand, the rate of production can be too slow at low temperatures. Therefore, it has been found that at 400°C, a substantial yield and production rate could be achieved. A higher temperature than this will result in a lower yield and higher operating costs. The latter comes from investment in expensive reactors that can withstand the higher temperature. In addition, a large amount of energy from the burning of fuels is continuously required to maintain the high temperature.

- SO_2 to O_2 is at 1:1 by volume
  Reason for choice: Since $V \propto n$ (Avogadro's Law), by introducing a volume ratio of 1:1 for SO_2 and O_2, this means that O_2 is in excess in accordance to the reaction stoichiometry.

  $$2SO_2(g) + O_2(g) \rightleftharpoons 2SO_3(g).$$

As oxygen is in excess, according to Le Chatelier's Principle, the position of equilibrium would shift toward the production of more SO_3, while trying to remove the excess O_2. This would help to produce more SO_3.

- Pressure: about 1 atm
  Reason for choice: Increasing the pressure of the reaction vessel will favor the forward reaction since the forward reaction will decrease the overall number of gaseous molecules and hence the pressure. This will increase the yield of $SO_3(g)$. However, operating at very high pressures will increase production costs, so an atmospheric pressure of 1 atm is used.

- The use of catalyst: vanadium(V) oxide (V_2O_5)
  Reason for choice: Although the use of a catalyst does not affect the yield of $SO_3(g)$, it will speed up the production rate and hence the time to reach equilibrium would be shorter.

Q How would you use kinetics to explain the effect of increasing pressure on the position of equilibrium?

A: According to Le Chatelier's Principle, when the total pressure of the system is increased by compression, concentrations of all species (both reactants and products) also increase as concentration is equal to $\frac{\text{amount in moles}}{\text{volume}}$. The system will attempt to decrease the overall pressure by favoring the reaction that decreases the overall number of gaseous molecules, i.e., the forward reaction is favored here. The equilibrium position will shift to the right. More $SO_2(g)$ and $O_2(g)$ will react to form $SO_3(g)$ until a new equilibrium is established.

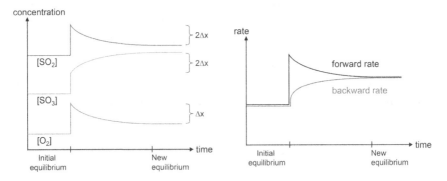

(d)   The sulfur trioxide produced above is dissolved in concentrated sulfuric acid to form oleum. Write the balanced chemical equation for this reaction.

Explanation:

$$SO_3(g) + H_2SO_4(l) \rightarrow H_2S_2O_7(l).$$

(e)   Explain why it is not conducive for sulfur trioxide to be dissolved in water to make sulfuric acid.

Explanation:

The direct dissolution of sulfur trioxide in water is highly exothermic and thus produces a lot of acid mist. So instead, the sulfur trioxide is dissolved in concentrated sulfuric acid to produce fuming sulfuric acid or oleum, which is then further diluted to produce concentrated sulfuric acid:

$$H_2S_2O_7(l) + H_2O(l) \rightarrow 2H_2SO_4(l).$$

(f)   Rust is a basic oxide consisting of Fe_2O_3. One use of sulfuric acid is to remove rust from steel sheets needed to make car bodies. Write the balanced chemical equation for the reaction between the surface layer of rust and dilute sulfuric acid.

Explanation:

$$Fe_2O_3(s) + 3H_2SO_4(aq) \rightarrow Fe_2(SO_4)_3(aq) + 3H_2O(l).$$

Do you know?

— Based on the Brønsted–Lowry Theory,

an acid is a proton donor;

a base is a proton acceptor; and

an acid–base reaction involves the acid transferring a proton to the base.

In the above example, the base is the O^{2-} ion of the ionic $Fe_2O_3(s)$ while the acid is the H^+ from the $H_2SO_4(aq)$. Thus, the water that is formed during the acid–base reaction is actually due to the following reaction:

$$2H^+ + O^{2-} \rightarrow H_2O.$$

(g) Sulfur dioxide is a reducing agent. Describe a test that can be used to show that sulfur dioxide is a reducing agent.

Explanation:

To test whether sulfur dioxide is a reducing agent, we need an oxidizing agent to react with the reducing agent. Acidified $KMnO_4$ can be used as the oxidizing agent. If the purple MnO_4^- solution turns colorless, then MnO_4^- has been reduced to Mn^{2+}. This means that the unknown substance has caused the reduction and hence contains a reducing agent.

	MnO_4^- manganate(VII) ion	Mn^{2+} manganese(II) ion
Oxidation state of Mn	+7	+2
Color	purple	colorless

In addition, we can also use acidified $K_2Cr_2O_7$. If the orange $Cr_2O_7^{2-}$ in the solution turns green, then $Cr_2O_7^{2-}$ has been reduced to Cr^{3+}. This means that the unknown substance has caused the reduction and hence contains a reducing agent.

	$Cr_2O_7^{2-}$ dichromate(VI) ion	Cr^{3+} chromium(III) ion
Oxidation state of Cr	+6	+3
Color	orange	green

Do you know?

— A redox reaction involves *electron transfer* between a pair of substances. In the process of electron transfer, one species will *lose* electron(s) to the other, which will *gain* the electron(s).

- The species that lose electron(s) is said to be oxidized, i.e., it undergoes oxidation. This species is known as the *reducing agent* or a *reductant*.
- The species that accepts the electron(s) is said to be reduced, i.e., it undergoes reduction. This species is known as the *oxidizing agent* or *oxidant*.

(h) Sulfur dioxide is also a bleaching agent and it bleaches by reduction. It is used to whiten wood pulp in the manufacture of paper. The bleaching effect is only temporary as the white paper will slowly turn yellowish-brown again. Explain.

Explanation:

The white paper will slowly turn yellowish-brown again because the oxygen in the air oxidizes the whitened wood pulp back to its original color before being reduced by sulfur dioxide.

(i) Draw the dot-and-cross diagrams for sulfur dioxide and sulfur trioxide.

Explanation:

$$\ddot{O}:\ddot{S}:\ddot{O} \qquad \ddot{O}:\ddot{S}:\ddot{O}$$

---

3. Ammonia produced from the Haber Process is used in the manufacturing of nitric(V) acid, HNO_3, via a two-step reaction:

Step 1: $4NH_3(g) + 5O_2(g) \rightleftharpoons 4NO(g) + 6H_2O(g)$,
$\Delta H = -950$ kJ mol^{-1}; and

Step 2: $4NO(g) + 2H_2O(g) + 3O_2(g) \rightarrow 4HNO_3(g)$.

(a) State the essential conditions used in the Haber Process for the manufacture of ammonia. Write the equation for the reaction that occurs in this process.

Explanation:

$$N_2(g) + 3H_2(g) \rightleftharpoons 2NH_3(g), \quad \Delta H = -76 \text{ kJ mol}^{-1}.$$

To maximize the yield of $NH_3(g)$ and minimize production costs, the following conditions are applied to the production process:

• Temperature: 450°C
Reason for choice: Lowering the temperature of the reaction vessel will favor the forward reaction since it is exothermic in nature. Hence, a higher yield of $NH_3(g)$ can be obtained.

On the other hand, the rate of production can be too slow at low temperatures. Therefore, it has been found that at 450°C, a substantial yield and production rate could be achieved. A higher temperature than this will result in a lower yield and higher operating costs. The latter comes from the investment in expensive reactors that can withstand the higher temperature. In addition, a large amount of energy from the burning of fuels is continuously required to maintain the high temperature.

- Pressure: about 200 atm
  Reason for choice: Increasing the pressure of the reaction vessel will favor the forward reaction since the forward reaction will decrease the overall number of gaseous molecules and hence the pressure. This will increase the yield of $NH_3(g)$. However, operating at very high pressures will increase production costs, so a moderate pressure of 200 atm is used.

- The use of catalyst: finely divided iron catalyst with aluminum oxide as the promoter
  Reason for use: Although the use of a catalyst does not affect the yield of $NH_3(g)$, it will speed up the production rate as the time that is needed to reach equilibrium would be shorter.

(b) Explain how the rate of reaction for Step 1 would change when

   (i) the pressure is increased by increasing the amount of ammonia molecules;

Explanation:

If the pressure is increased by increasing the amount of ammonia molecules, the frequency of collisions between the ammonia molecules with the other reactants would increase. Thus, the rate of the forward reaction for Step 1 would increase.

Q Would the increase in rate of the reaction for Step 1 have any impact on the position of equilibrium for Step 1?

A: Yes! As the rate for the forward reaction is greater than the rate for the backward reaction, the position of equilibrium would shift to the right as shown below:

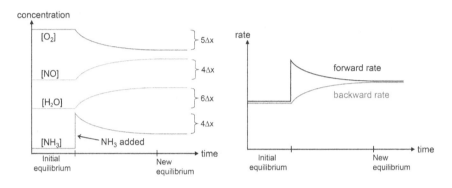

Q Why isn't there a complete removal of all the ammonia that has been added into the system?

A: When a system at equilibrium is subjected to a change, there is no way for the system to completely remove the change! Think logically, if all the added ammonia has been completely removed, then the amount of O_2 is going to be much smaller than in the beginning because the additional added NH_3 would use up the O_2. But at the point of the new equilibrium, the amounts of NO and H_2O are much greater than those in the beginning. So, how can both the forward and backward rates be the same at the new equilibrium position? In addition, at the new equilibrium, with a smaller amount of O_2 but same amount of NH_3, the forward rate has to be lower than that for the initial equilibrium before the change has taken place.

(ii) the pressure is increased by increasing the amount of nitrogen monoxide molecules;

Explanation:

If the pressure is increased by increasing the amount of nitrogen monoxide molecules, the frequency of collisions between the nitrogen monoxide molecules with the other reactants would increase. Thus, the rate of the backward reaction for Step 1 would increase, which means that the rate of the forward reaction would also increase at the new equilibrium position as shown below:

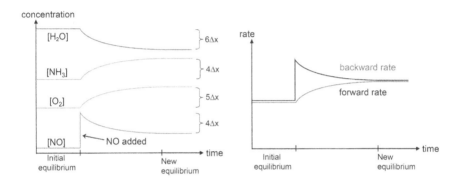

> (iii) the pressure is increased by reducing the volume of the container; and

Explanation:

If the pressure is increased by reducing the volume of the container, the concentrations of all the different types of particles would increase. The frequency of collisions for the particles in the forward reaction would increase, and similarly for the backward reaction. Hence, both the rates of reaction for the forward and backward directions would increase.

Q Since both the rates of reaction for the forward and backward directions would increase, do they both increase to the same extent?

A: No, in this case. Since the forward reaction involves the collision of nine particles while the backward reaction involves the collision of 10 particles, there is a greater proportional increase in the rate of the backward reaction than that for the forward reaction. Thus, as time passes, the position of equilibrium is going to shift to the left. As predicted by Le Chatelier's Principle, a decrease in the increase in pressure that is brought about by the reduction of the volume of the container, will produce a smaller number of particles.

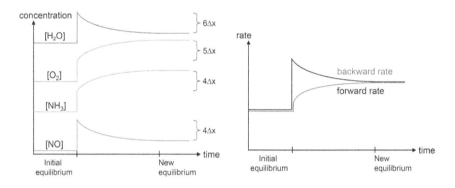

(iv) decreasing temperature.

Explanation:

If the temperature is decreased, the particles would have less kinetic energy. As a result, there would be a smaller amount of particles possessing kinetic energies that are greater than the activation energy. Hence, the frequency of effective collisions would decrease, causing a decrease in both the forward and backward rates.

Q Since Step 1 is an exothermic reaction, does the decrease in temperature cause the position of equilibrium to shift to the right?

A: Yes, indeed. When the temperature of the system is decreased, the rates of both the forward and backward reactions would decrease. But the percentage decrease is greater for the reaction that has the greater activation energy, i.e., the backward reaction for Step 1. So, as time passes, the position of equilibrium would shift to the *right*. This explains how the system "supplies" more heat to counteract the decrease in temperature. This is the rationale behind using Le Chatelier's Principle to predict the shift in the position of equilibrium.

(c) Explain whether you would get more nitrogen monoxide for each of the cases in part (b).

Explanation:

(i) The amount of NO at the new equilibrium is greater than in the old equilibrium as the position of equilibrium has shifted to the right.

(ii) The amount of NO at the new equilibrium is greater than in the old equilibrium, although the position of equilibrium has shifted to the left. This is because not all of the additional amount of NO that has been added into the system, are completely being removed. And certainly, some of the added NO would have been "wasted" because they were converted to the reactants.

(iii) The amount of NO at the new equilibrium is less than in the old equilibrium as the position of equilibrium has shifted to the left.

(iv) The amount of NO at the new equilibrium is greater than in the old equilibrium as the position of equilibrium has shifted to the right.

(d) During the reaction, ammonia and oxygen are passed through a powdered catalyst.

(i) Explain with a labeled diagram how a catalyst speeds up the rate of the reaction.

Explanation:

A catalyzed reaction proceeds via an *alternative pathway* of a lower activation energy, E_a, in contrast to the uncatalyzed reaction. The energy profile diagram for a catalyzed and an uncatalyzed reaction is shown below:

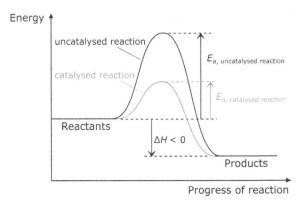

As a result of the lowered activation energy for the catalyzed reaction, there is now a greater proportion of reacting particles having sufficient kinetic energy greater than the *lowered E'_a*. This is indicated by the larger shaded area in the Maxwell–Boltzmann distribution diagram that is shown below:

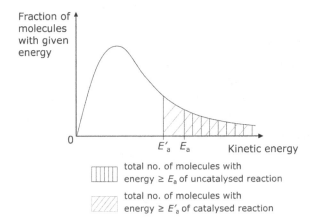

Thus, having a greater number of reactant particles with a kinetic energy greater than or equal to the lowered activation energy (E_a') would result in an increase in the *frequency of effective collisions*. This would thus lead to an increase in the reaction rate.

(ii) Does the catalyst change the ΔH for the reaction? Explain.

Explanation:

The catalyst does not change the ΔH for the reaction. This is because a catalyst lowers the activation energies of both the forward and backward reactions to the same extent. This, in turn, would lead to the rates of both the forward and backward reactions to be increased to the same extent.

(iii) Explain why the catalyst becomes hot during the reaction.

Explanation:

As the reaction is exothermic in nature, the heat energy that is evolved would cause the catalyst to become hot.

(iv) Explain why the catalyst is used in the form of a powder.

Explanation:

A powdered catalyst has a greater surface area than when the catalyst is in a lump. This allows more reactants to adsorb on the greater surface area for reaction, thus increasing the rate of reaction.

(v) If a better catalyst is used to replace the original one, would this affect the yield of nitrogen monoxide? Explain your answer.

Explanation:

The use of a catalyst does not change the position of equilibrium; hence, the yield of the product remains unchanged. Even if a better catalyst is used, the yield would still be the same as that of a non-catalyzed reaction.

Q So, the better catalyst does not have any impact on the reaction at all?

A: Not really. A better catalyst would make the reaction reach the equilibrium position faster. Hence, it saves time.

(e) Use the two equations to construct an overall equation for the conversion of ammonia to nitric acid.

Explanation:

Step 1: $4NH_3(g) + 5O_2(g) \rightleftharpoons 4NO(g) + 6H_2O(g).$
Step 2: $4NO(g) + 2H_2O(g) + 3O_2(g) \rightarrow 4HNO_3(g).$
Overall equation: $4NH_3(g) + 8O_2(g) \rightarrow 4HNO_3(g) + 4H_2O(g).$

(f) If all the ammonia is converted to nitric acid, calculate the mass of nitric acid obtained when 1 dm^3 of ammonia is used at s.t.p.

Explanation:

At s.t.p., one mole of gas occupies a volume of $22.7 \, dm^3$.

Amount of NH_3 in $1 \, dm^3 = \dfrac{1}{22.7} = 4.41 \times 10^{-2} \, mol$.

Four moles of NH_3 give four moles of HNO_3; 4.46×10^{-2} mol gives 4.41×10^{-2} mol of HNO_3.

Molar mass of $HNO_3 = 1.0 + 14.0 + 3(16.0) = 63.0 \, g \, mol^{-1}$.

Mass of nitric acid obtained $= 4.41 \times 10^{-2} \times 63.0 = 2.78 \, g$.

(g) It is possible to monitor the rate of the reaction for Step 1 by following pH changes during the reaction. Samples of gas were taken from the reaction vessel at regular intervals and bubbled through water to form a solution. The pH of each solution was measured. Explain why the measured pH changes during the reaction.

Explanation:

Ammonia is a basic gas that hydrolyzes in water, producing $OH^-(aq)$ as follows:

$$NH_3(g) + H_2O(l) \rightleftharpoons NH_4^+(aq) + OH^-(aq).$$

As the reaction proceeds, the ammonia is consumed. Hence, a smaller amount of ammonia would mean that a smaller amount of $OH^-(aq)$ is produced, causing the measured pH to change during the reaction.

 Since the reaction produces HNO_3, wouldn't the HNO_3 react with the basic NH_3 and thus affect our measurement of the rate of reaction?

A: The rate of the formation of HNO_3 is equivalent to the rate in which the NH_3 is converted to HNO_3. Thus, the actual rate of "disappearance" of NH_3 is a "compounded" rate, i.e., the sum of the rate of converting the NH_3 to HNO_3 and the rate of reaction of the NH_3 with the HNO_3 produced. Hence, although we are measuring the "compounded" rate, it would not affect the shape of the graph that we would obtain, so don't worry too much.

(h) Suggest another method that can be used to measure the rate of the reaction.

Explanation:

As the reaction proceeds, the pressure of the system changes as the number of particles decreases (12 gaseous particles become eight gaseous particles since $4NH_3(g) + 8O_2(g) \rightarrow 4HNO_3(g) + 4H_2O(g)$). Thus, by monitoring the pressure, it can be used to measure the rate of the reaction.

 If the HNO_3 is in the aqueous state, it should not affect the pressure of the system, right?

A: You are right in saying that the aqueous HNO_3 would not affect the pressure of the system, which originates in the collision of the gaseous particles with the wall of the container.

(i) Draw the dot-and-cross diagram for ammonia.

Explanation:

$$\begin{array}{c} H \\ H \overset{\bullet}{\underset{\bullet\bullet}{\overset{\times}{N}}} H \end{array}$$

4. This question concerns the reaction

$$H_2(g) + I_2(g) \rightleftharpoons 2HI(g),$$

which, even at high temperatures, is slow.

(a) (i) Given the following information,

Covalent bond	Bond energy in kJ mol^{-1}
H–H	436
I–I	151
H–I	299

calculate ΔH for the reaction between hydrogen and iodine.

Explanation:

$$H–H + I–I \rightarrow 2H–I.$$

Total energy absorbed during bond breaking = BE(H–H) + BE(I–I)

$$= 436 + 151$$

$$= 587 \text{ kJ mol}^{-1}.$$

Total energy released during bond forming $= 2 \times$ BE(H–I)

$$= 598 \text{ kJ mol}^{-1}.$$

ΔH = (Sum of total energy for bond breaking)
  – (Sum of total energy for bond making).

Hence, $\Delta H = 587 - 598 = -1 \text{ kJ mol}^{-1}$.

(ii) Sketch an energy level diagram for this reaction.

Explanation:

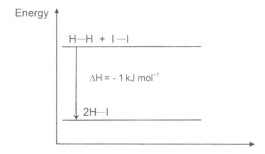

(b) Indicate on your energy profile diagram
   (i) ΔH for the reaction;
   (ii) the activation energy for the forward reaction ($E_{a(f)}$); and
   (iii) the activation energy for the reverse reaction ($E_{a(b)}$).

Explanation:

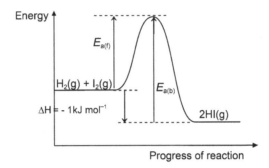

The graph shows: Energy (vertical axis) vs Progress of reaction (horizontal axis). $E_{a(f)}$ is the activation energy of the forward reaction. The starting level is labelled $H_2(g) + I_2(g)$, with $\Delta H = -1\text{kJ mol}^{-1}$. $E_{a(b)}$ is the activation energy of the backward reaction, and the product level is labelled $2HI(g)$.

(c) In an analogous reaction for the formation of hydrogen chloride,

$$H_2(g) + Cl_2(g) \rightarrow 2HCl(g),$$

suggest how you would expect the activation energy of the forward reaction to be like when it is compared with that shown for the formation of HI. Give a reason for your answer.

Explanation:

As the Cl–Cl bond is stronger than the I–I bond, more energy is needed to break the Cl–Cl bond. Thus, we would expect the activation energy of the forward reaction to be more endothermic.

 So, the activation energy of the $2HCl(g) \rightarrow H_2(g) + Cl_2(g)$ reaction would also be more endothermic than the decomposition of HI(g)?

A: Yes, indeed. This is because more energy is needed to break the stronger H–Cl bond as compared to the H–I bond.

(d) The reaction for the formation of HI does not go to completion but reaches equilibrium. In an experiment to establish the equilibrium concentration, the reaction was allowed to reach equilibrium at 723 K and then quenched by addition to a known, large volume of cold water. The concentration of iodine in this solution was then determined by titration with standard sodium thiosulfate solution.

(i) Explain what is meant by the word *quench* and why quenching is necessary.

Explanation:

The purpose of quenching is to slow down the reaction. This is made possible here by adding water, i.e., increase the volume of the solution. In a larger-volume condition, the reacting particles are farther apart and are less likely to collide with each other.

Q Why is cold water preferred for quenching?

A: Cold water lowers the temperature and hence the rate of the reaction. This helps to ensure that the concentrations of the species in the system are as it is at the temperature of measurement. In this case here, the temperature is 723K. In short, the sudden drop in temperature "freeze" the position of equilibrium at 723K and enable the composition of the system to be analyzed at a lower temperature.

(ii) Write an equation between sodium thiosulfate and iodine.

Explanation:

$$I_2(aq) + 2S_2O_3^{2-} \rightarrow 2I^-(aq) + S_4O_6^{2-}(aq).$$

(iii) What indicator would you use? Give the color change at the endpoint.

Explanation:

Starch solution is added to the solution when it turns pale yellow. A blue-black coloration is observed. The endpoint is reached when the blue-black coloration decolorizes.

Do you know?

— The blue-black coloration is due to the formation of a starch–iodine complex. The starch solution cannot be added too early when the concentration of the iodine is still high as too much of the complex would form. The insolubility of the starch–iodine complex may prevent some of the iodine from reacting with the thiosulfate as they are being embedded within the starch matrix.

(iv) In this titration and in titrations involving potassium manganate(VII), a color change occurs during reaction. Why is an indicator usually added in iodine/thiosulfate titrations but not in titrations involving potassium manganate(VII)?

Explanation:

When the purple potassium manganate(VII) is used during titration, the purple MnO_4^- ion is converted to the near colorless Mn^{2+}. Hence, the endpoint can be easily distinguished when one drop of excess purple potassium manganate(VII) causes the solution to turn pink.

But for iodine–thiosulfate titration, toward the endpoint, the solution turns pale yellow. Hence, it is not easy to detect the color transition from pale yellow to colorless. Therefore, by creating a blue-black coloration, we can prominently detect the color transition from blue-black to colorless.

5. Methanol is produced by reacting carbon monoxide and hydrogen as shown in the following equation:

$$CO(g) + 2H_2(g) \rightleftharpoons CH_3OH(g), \quad \Delta H = -90 \text{ kJ mol}^{-1}.$$

The reaction is carried out at high pressure and the gases are passed through a mixture of zinc oxide and chromium oxide maintained at 400°C. These conditions are chosen to give a good yield of methanol at a satisfactory rate.

(a) Draw an energy profile for the above reaction.

Explanation:

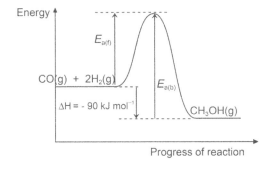

(b) Explain why increasing the pressure gives a better yield, and give one disadvantage of using high pressure in an industrial plant.

Explanation:

When pressure is increased, the concentrations of the different types of particles in the system are increased. According to Le Chatelier's Principle, the system would respond in such a way so as to decrease the amount of particles. Since the forward reaction produces a smaller amount of particles than the backward reaction, the position of equilibrium would shift to the right.

Operating at a high pressure in an industrial plant would increase the operating cost since a more expensive reactor that can withstand high pressures has to be used.

(c) The value of ΔH is negative. What does this tell you about the relative stabilities of the reactants and products?

Explanation:

The negative value of ΔH indicates that the products are at a lower energy level than the reactants. Hence, the products are more stable than the reactants. This would also mean that the bonds in the products are stronger than the bonds in the reactants.

(d) State, with a reason, whether a high or low temperature will give a better yield of methanol at equilibrium.

Explanation:

Since the reaction is exothermic in nature, a low temperature will give a better yield of methanol at equilibrium. According to Le Chatelier's Principle, at low temperature, the system would response in such a way so as to produce more heat energy. Since the forward reaction is exothermic in nature, the position of equilibrium would shift to the right.

(e) With reference to your answer in part (d), explain why a temperature of 400°C is used in practice.

Explanation:

Lowering the temperature of the reaction vessel will favor the forward reaction since it is exothermic in nature, and hence, a higher yield of methanol can be obtained.

On the other hand, the rate of production can be too slow at low temperatures. Therefore, it has been found that at 400°C, a substantial yield and production rate could be achieved. A higher temperature than this will result in a lower yield and higher operating costs. The latter comes from the investment in expensive reactors that can withstand the higher temperature. In addition, a large amount of energy from the burning of fuels is continuously required to maintain the high temperature.

(f) Suggest why the gases are passed through a mixture of zinc oxide and chromium oxide.

Explanation:

The mixture of zinc oxide and chromium oxide is the catalyst. It speeds up the rate to achieve equilibrium and shortens the time to obtain the product.

(g) Calculate the maximum mass of methanol that can be obtained from 20 kg of hydrogen.

Explanation:

$$CO(g) + 2H_2(g) \rightleftharpoons CH_3OH(g).$$

Amount of H_2 in 20 kg $= \dfrac{20,000}{2} = 10,000$ mol.

Amount of CH_3OH produced $= \frac{1}{2} \times 10{,}000 = 5000$ mol.

Maximum mass of $CH_3OH = 5000 \times (12.0 + 3.0 + 16.0 + 1.0) = 160$ kg.

> (h) Give the dot-and-cross diagram of methanol, where there is a C–O single bond in the molecule.

Explanation

CHAPTER 8

ACIDS, BASES, AND SALTS

1. Write the ionic equations, with state symbols, for the following reactions:

   (a) Barium nitrate solution reacts with sodium sulfate solution to form an insoluble salt and sodium nitrate solution.

Explanation:

$$Ba^{2+}(aq) + SO_4^{2-}(aq) \rightarrow BaSO_4(s).$$

Q   Why are some ionic compounds soluble in water while others are not?

A: For ionic compounds that are soluble in water, we can say that in *general*, it is because the attractive forces that are formed between the ions and water molecules are able to provide a sufficient amount of energy to break up the ionic lattice. For those that are insoluble, it is probably due to the insufficient amount of energy supplied to overcome the greater strength of ionic bonds. Thus, we can classify salts into *soluble* salts and *insoluble* salts:

Soluble salts	Insoluble salts
• Salts containing Na^+, K^+, and NH_4^+ as cation are soluble.	
• Salts containing NO_3^- as the anion are soluble.	
• Salts containing Cl^-, Br^-, and I^- as the anion are soluble EXCEPT	• $PbCl_2$, $PbBr_2$, PbI_2 • $AgCl$, $AgBr$, AgI
• Salts containing SO_4^{2-} as the anion are soluble EXCEPT	• $PbSO_4$, $BaSO_4$ and $CaSO_4$ (sparingly soluble)
• Na_2CO_3, K_2CO_3, $(NH_4)_2CO_3$.	• Salts containing CO_3^{2-} as the anion are insoluble EXCEPT

Do you know?

— To prepare an insoluble salt:

Step 1:   Mix two solutions together.

$$Ba(NO_3)_2(aq) + Na_2SO_4(aq) \rightarrow BaSO_4(s) + 2NaNO_3(aq).$$

Step 2:   Filter the mixture and collect the residue.

Step 3:   Wash the residue with deionized water.

Step 4:   Dry it and you get the salt crystals!

(b) Magnesium reacts with dilute hydrochloric acid to form a soluble salt and hydrogen.

Explanation:

$$Mg(s) + 2H^+(aq) \rightarrow Mg^{2+}(aq) + H_2(g).$$

Do you know?

— There are three important characteristics that an acid may display:

- An acid reacts with a metal to give off hydrogen gas.
- An acid reacts with a carbonate/hydrogencarbonate to give off carbon dioxide gas.
- An acid reacts with base to give a salt and water.

— But not all acids would display all the above three characteristics simultaneously. Some acids may just display one out of the three. For example, water reacts with sodium metal to give off hydrogen gas but does not react with carbonates or with a base.

Q How do you prepare a soluble salt?

A: To prepare a soluble salt:

Step 1: Get the salt solution.

- Acid reacting with excess metal:

$Zn(s) + 2HCl(aq) \rightarrow ZnCl_2(aq) + H_2(g)$ + unreacted metal.

- Acid reacting with excess insoluble base:

$Zn(OH)_2(s) + 2HCl(aq)$
$\rightarrow ZnCl_2(aq) + 2H_2O(l)$ + unreacted base.

- Acid reacting with excess insoluble carbonate:

$ZnCO_3(s) + 2HCl(aq) \rightarrow ZnCl_2(aq) + CO_2(g)$
$+ H_2O(l)$ + unreacted carbonate.

Step 2: Filter the mixture and collect the filtrate.

Step 3: Concentrate the filtrate through evaporation.

Step 4: Crystallize out the salt by cooling the hot solution.

Step 5: Filter and you get the salt crystals!

  Why is excess metal, excess base, excess carbonate used when preparing the salt?

A: As the metal, base, and carbonate used are insoluble, when we use these substances in excess in the preparation of the salt, we ensure that all the acid is completely reacted. Otherwise later during the crystallization of the salt, the salt might be contaminated with the leftover acid.

Q   How would you crystallize the salt out?

A: If we have a mixture where one of the components is soluble and we would like to have the water of crystallization in the solid form:

Step 1:   Dissolve the mixture with excess water in a beaker.

Step 2:   Filter the mixture with a filter funnel and collect the filtrate.

Step 3:   Concentrate the filtrate by evaporating most of the water off.

Step 4:   Let the saturated solution cool down slowly.

Step 5:   Put in a small crystal of the pure salt as a "seed." As the temperature decreases, the solubility of the salt also decreases. The salt would then deposit on the seed of crystal.

Concentrated solution
cooling from high temperature
to low temperature

Seeding crystal
tied to a string

(c) Zinc reacts with aqueous copper(II) sulfate to form aqueous zinc(II) sulfate and copper metal.

Explanation:

$$Zn(s) + Cu^{2+}(aq) \rightarrow Zn^{2+}(aq) + Cu(s).$$

Do you know?

— The above reaction is a displacement reaction. In fact, it is a *redox* (*red*uction–*ox*idation) reaction. The Zn metal becomes Zn^{2+} by losing electrons, while the Cu^{2+} becomes Cu by gaining electrons. The Zn metal is said to have undergone oxidation (i.e., being oxidized), while the Cu^{2+} undergoes reduction (i.e., being reduced). There is a transfer of electrons from the Zn atom to the Cu^{2+} ion!

(d) Calcium carbonate reacts with sulfuric acid to form carbon dioxide, salt, and water.

Explanation:

$$CaCO_3(s) + 2H^+(aq) + SO_4^{2-}(aq) \rightarrow CaSO_4(s) + CO_2(g) + H_2O(l).$$

(e) Calcium carbonate reacts with nitric acid to form carbon dioxide, salt, and water.

Explanation:

$$CaCO_3(s) + 2H^+(aq) \rightarrow Ca^{2+}(aq) + CO_2(g) + H_2O(l).$$

Do you know?

— CO_2 is not very soluble in water and at the same time, if there is a large evolution of the gas, the gas escapes from a liquid, causing a bubbling effect. This bubbling effect is known as *effervescence*. Do not mistake effervescence for the bubbling that you see during boiling. The bubbles that you see during boiling are actually bubbles of water vapor.

(f) Aluminum oxide reacts with sulfuric acid to form salt and water.

Explanation:

$$Al_2O_3(s) + 6H^+(aq) \rightarrow 2Al^{3+}(aq) + 3H_2O(l).$$

2. Silicon(IV) oxide and magnesium oxide react to form magnesium silicate.
   (a) Write the balanced chemical equation for the reaction.

Explanation:

$$SiO_2(s) + MgO(s) \rightarrow MgSiO_3(s).$$

Do you know?

— Magnesium silicate is a major component of talcum powder, which is widely used in the cosmetic industry, and also in baby powder. It is likewise often used in basketball or weightlifting to keep the player's hands dry.

(b) Explain why these two oxides can react together.

Explanation:

MgO is a basic oxide while SiO_2 is acidic in nature. Thus, the reaction between the two oxides is in fact an acid–base reaction.

Q MgO consists of Mg^{2+} and O^{2-} ions; how does the O^{2-} ion get "incorporated" into SiO_2?

A: The SiO_2 has a giant covalent/macromolecular lattice structure in which each Si atom is covalently bonded to four other O atoms, and each O atom is in turn bonded to two Si atoms.

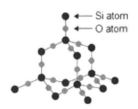

Since the Si atom is bonded to four highly electronegative O atoms, the Si atom is highly electron-deficient as the shared covalent electrons are increasingly attracted toward the O atoms. As a result, the highly electron-rich O^{2-} ion can easily be attracted to the electron-deficient Si atom. It is simply another example of electrostatic attraction between oppositely charged particles.

Do you know?

— There are four types of oxides: basic oxides, acidic oxides, neutral oxides, and amphoteric oxides.

- Basic oxides are formed from the reaction of metals with oxygen, for example, Na_2O, MgO, and CaO. Basic oxides can react with an acid or even with an acidic oxide:

$$Na_2O(s) + CO_2(g) \rightarrow Na_2CO_3(s).$$

- Acidic oxides result from the reaction of non-metals, such as CO_2 and SO_2, with oxygen. It reacts with a base to give a salt and water:

$$CO_2(g) + Ca(OH)_2(aq) \rightarrow CaCO_3(s) + H_2O(l).$$

- Neutral oxides neither react with acids nor bases. An example is CO.
- Amphoteric oxides react with both acids and alkalis. It is formed from the reaction of a metal with oxygen. Examples are Al_2O_3, ZnO, and PbO. The oxide ion component, O^{2-}, enables the amphoteric oxide to react with an acid. Due to the high *charge density* of the cation, the metal cation component reacts with a base as follows:

$$Al_2O_3(aq) + 6HCl(aq) \rightarrow 2AlCl_3(aq) + 3H_2O(l)$$

$$Al_2O_3(s) + 2NaOH(aq) \rightarrow 2NaAlO_2(aq) + H_2O(l)$$
sodium aluminate

$$ZnO(s) + 2HCl(aq) \rightarrow ZnCl_2(aq) + H_2O(l)$$

$$ZnO(s) + 2NaOH(aq) \rightarrow Na_2ZnO_2(aq) + H_2O(l).$$
sodium zincate

Q Why does the high charge density of the cation enable it to react with the base?

A: If the charge density of the positive cation is very high, it would be strong enough to attract the base, OH^-, which has an opposite charge as compared to the cation. And if the attractive force is so strong, then the two oppositely charged particles can form another particle, such as AlO_2^- and ZnO_2^{2-}.

---

3. Lactic acid is a weak monobasic acid that is commonly found in sour milk. A $0.10\,mol\,dm^{-3}$ solution of lactic acid has a pH of 2.43.

   (a) Explain the meaning of the terms *weak* and *monobasic*.

Explanation:

A weak acid undergoes partial dissociation in aqueous solution.

The *basicity of an acid* refers to the number of moles of *hydroxide ions* that is needed to completely neutralize *one mole* of the acid molecule.

Do you know?

— According to Brønsted–Lowry definition, an acid is a proton donor while a base is a proton acceptor. This would mean that in order for an acid–base reaction to occur, both the acid and base must coexist to facilitate proton transfer. But what happens to the proton after it is accepted by the base? The proton forms a dative covalent bond/-coordinate bond with the base:

$$H-\underset{H}{\overset{\cdot\cdot}{N}}-H \ + \ H^+ \ \longrightarrow \ \left[H-\underset{H}{\overset{H}{N}}-H \right]^+$$

— A *strong* acid *completely dissociates* in aqueous solution. That is, one mole of HCl will provide one mole of H_3O^+ and one mole of Cl^- upon

(Continued)

(Continued)

dissociation in water, i.e., [HCl] = [H_3O^+]. We use a single arrow "→" for the dissociation equation:

$$HCl(aq) + H_2O(l) \rightarrow H_3O^+(aq) + Cl^-(aq).$$

— As a weak acid does not fully dissociate in water, one mole of CH_3COOH will provide less than one mole of H_3O^+ and CH_3COO^- *each*, upon partial dissociation in water, i.e., [CH_3COOH]$_{initial}$ > [H_3O^+]$_{equilibrium}$. We use a double arrow "⇌" for the dissociation equation:

$$CH_3COOH(aq) + H_2O(l) \rightleftharpoons CH_3COO^-(aq) + H_3O^+(aq).$$

A weak acid dissociates to a lesser extent than a strong acid because the dissociation of a weak acid is energetically less favorable than that for a strong acid. A weak acid needs to absorb more energy in order to fully dissociate!

Q Can we simply write the equation as $HCl(aq) \rightarrow Cl^-(aq) + H^+(aq)$ to indicate that HCl is acidic?

A: For the purpose of simplicity, yes. But it is more meaningful to write the equation as: $HCl(aq) + H_2O(l) \rightarrow Cl^-(aq) + H_3O^+(aq)$. This is because the H–Cl bond does not automatically break up when a HCl molecule "plunges" into the water. The "loss" of a proton from a H–Cl molecule is mediated by the lone pair of electrons from a water molecule. One can actually visualize the lone pair of electrons "extracting" the proton. In addition, hydrogen ions do not exist in solution. A H^+ ion has such a high charge density that it is actually bonded to at least one water molecule when in aqueous solution, i.e., H^+ binds with H_2O to form H_3O^+ (hydronium ion).

Without the water molecule, the HCl cannot function as an acid! It is common to find the symbol "H^+(aq)" used in many texts for simplicity's sake, including this one. It is okay to use it but we must bear in mind that when we write "H^+(aq)," we are actually referring to "H_3O^+(aq)."

> **Q**   What is a lone pair of electrons?

A: A lone pair of electrons is a pair of electrons that is in the valence shell that is not involved in covalent bonding.

Do you know?

— Acids are classified as monoprotic, diprotic, or triprotic, depending on the number of H atoms in a molecule that are able to ionize to form H^+ ions. A monoprotic acid is a monobasic acid because one mole of the acid molecules needs one mole of hydroxide ions (OH^-) to fully neutralize it, and so on and so forth for dibasic and tribasic acids. Thus, the *basicity of an acid* refers to the number of moles of *hydroxide ions* that is needed to completely neutralize *one mole* of the acid molecule. It should not be confused with the term "basicity" alone, which refers to how alkaline the solution is!

— But not all hydrogen atoms in a molecule will ionize to form H^+ ions. For example, in ethanoic acid, only one hydrogen atom out of the four will ionize:

$$CH_3COOH \rightleftharpoons H^+ + CH_3COO^-.$$

— The basicity of an acid should not be confused with the strength of an acid. The number of hydrogen ions liberated per molecule of acid does not determine its strength. A dibasic or tribasic acid need not necessarily be a stronger acid than a monobasic acid. A good example is carbonic acid, H_2CO_3, which is a *weak dibasic* acid. On the other hand, both nitric acid and hydrochloric acid are strong monobasic acids.

(b) Calculate the concentration of hydrogen ions present in the above acid solution.

Explanation:

The pH of a solution is defined as negative $\log_{10}$ of the concentration of hydrogen ion in $mol\,dm^{-3}$

$$pH = -\log_{10} [H^+(aq)].$$

Thus, given pH = 2.43, the $[H^+(aq)] = 10^{-2.43} = 3.72 \times 10^{-3}\,mol\,dm^{-3}$.

Do you know?

— The pH scale value

- has no units;
- ranges from 0 to 14; and
- the greater the $[H^+]$, the smaller the pH value.

By knowing the $[H^+]$, we can calculate the pH of an acid, and vice versa for an alkali!

Q What if we are asked to calculate the pH of NaOH(aq)? NaOH(aq) is a base and it doesn't produce H^+ ions. So, how is it possible to assign a pH value to an aqueous solution of NaOH(aq) when we do not have a value for $[H^+]$?

A: Well, it turns out that for all aqueous solutions, be it acidic, neutral, or basic, there are both H^+ and OH^- ions present. In *pure water*, the water molecules actually undergo *auto-ionization*:

$$H_2O(l) + H_2O(l) \rightleftharpoons H_3O^+(aq) + OH^-(aq),$$

$[H_3O^+(aq)] = [OH^-(aq)] = 10^{-7}\,mol\,dm^{-3}$ at 25°C, i.e., the reason why the pH of pure water is 7! $[H_3O^+(aq)]$ varies depending on whether an acid or base is added to the pure water.

Depending on the types of solution, the concentrations of these two ions differ:

- for neutral solutions, $[H_3O^+(aq)] = [OH^-(aq)]$;
- for acidic solutions, $[H_3O^+(aq)] > [OH^-(aq)]$; and
- for basic solutions, $[H_3O^+(aq)] < [OH^-(aq)]$.

The product of $[H_3O^+(aq)][OH^-(aq)] = 10^{-14}\,mol^2\,dm^{-6}$, hence we can calculate the $[H_3O^+(aq)]$ simply by using $[H_3O^+(aq)] = 10^{-14}/[OH^-(aq)]$.

Many students assume that there is no OH^- ions in an acidic solution and vice versa. This is incorrect! Both the $H_3O^+(aq)$ and $OH^-(aq)$ ions are always present simultaneously in a solution!

(c) With reference to your answer in part (b), explain whether lactic acid is a strong or weak acid.

Explanation:

Lactic acid is a weak acid since the $[H^+(aq)] = 3.72 \times 10^{-3}\,mol\,dm^{-3} \neq 0.10\,mol\,dm^{-3}$. This shows that not all lactic acid molecules have completely dissociated.

(d) Calculate the volume of sodium hydroxide with a concentration of $0.10\,mol\,dm^{-3}$ that is needed to neutralize $30\,cm^3$ of the lactic acid solution.

Explanation:

Amount of lactic acid $= \dfrac{30}{1000} \times 0.10 = 3.0 \times 10^{-3}\,mol$.

Since lactic acid is monobasic, let it be HA. Thus, the ionic equation for the acid–base reaction is

$$HA(aq) + OH^-(aq) \rightarrow A^-(aq) + H_2O(l).$$

Amount of lactic acid = Amount of NaOH = 3.0×10^{-3} mol.

Volume of NaOH = $\dfrac{0.003}{0.10}$ = 0.03 dm^3 = 30.0 cm^3.

Do you know?

— Although a weak acid does not fully dissociate in water, the amount of a strong base, such as NaOH, required for complete neutralization is still the same as compared to one that is needed to neutralize a strong acid of the same concentration, same volume and same basicity.

Wait a minute, you say that the amount of a strong base, such as NaOH, required to neutralize a weak acid and a strong acid of the same concentration, same volume and same basicity, are the same. Is it because NaOH is too strong that it is *not able* to differentiate the strengths of the two acids?

A: Now, an acid such as CH$_3$COOH, is weak as compared to HCl when we use H$_2$O as a base to differentiate their relative strengths. But if a stronger base is used, such as NaOH, it cannot differentiate between the strengths of the two acids; the NaOH would fully react with these two acids. So, you are right! Hence, it is important to take note that "strong" or "weak" is a relative concept. As the saying goes, "one base's strong acid is another base's weak acid."

Q What is basicity of an acid?

A: Basicity of an acid refers to the number of moles of hydroxide ions that is needed to completely neutralize one mole of the acid molecules. For examples, HCl is a monobasic acid while H$_2$SO$_4$ is a dibasic acid. Similarly, we say that NaOH is a monoacidic base because one mole of NaOH needs one mole of H$^+$ ions for complete neutralization while Ca(OH)$_2$ is a diacidic base.

> (e) Suggest a suitable indicator for the above titration.

Explanation:

Since this is a weak acid–strong base titration, a suitable indicator would be phenolphthalein or bromothymol blue.

 Q What is the reason behind the choice of indicator for a weak acid–strong base titration?

A: Simple! What makes a weak acid weak in nature? The "unwillingness" to dissociate to provide a proton, H^+, right? So, after the weak acid has dissociated, do you see that there is a high "affinity" for the conjugate base to "grab" a proton back if possible? This accounts for the reversibility of the backward reaction ($CH_3COOH(aq) + H_2O(l) \rightleftharpoons CH_3COO^-(aq) + H_3O^+(aq)$). For a strong acid, such as HCl, you do not have the backward reaction occurring simply because the Cl^- ion that is formed has very low affinity to take "back" a H^+ ion! So, the dissociation of HCl is an irreversible reaction ($HCl(aq) + H_2O(l) \rightarrow Cl^-(aq) + H_3O^+(aq)$).

　Now, the endpoint pH of a strong base–weak acid titration is not at the neutral pH of 7, but it is in the alkaline range. The pH is greater than 7 because of basic hydrolysis that is brought about by the conjugate base (A^-) of the weak acid (HA):

$$A^-(aq) + H_2O(l) \rightleftharpoons HA(aq) + OH^-(aq).$$

Both phenolphthalein and bromothymol blue change color in the alkaline pH range. Hence, they are suitable indicators for the weak acid–strong base titration.

 Q So, does this mean that the endpoint pH of a weak base–strong acid titration would have a pH < 7?

A: Yes, indeed. The endpoint pH of a weak base–strong acid titration would have a pH < 7 because of acidic hydrolysis brought about by the conjugate acid (e.g., NH_4^+) of the weak base (e.g., NH_3), i.e., $NH_4^+(aq) + H_2O(l) \rightleftharpoons NH_3(aq) + H_3O^+(aq)$. Hence, the choice of indicator for different types of titration is important! The following shows three possible titration curves:

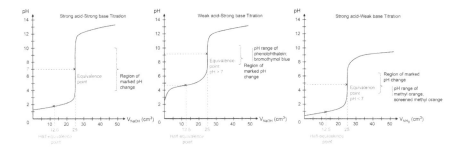

Different indicators change colour at different pH ranges; this is known as the working range of the indicator.

Q So, does that mean that if we use methyl orange as the indicator for a weak acid–strong base titration, the moment the methyl orange changes color (at working range of pH 3–4), it would mean that we have added a smaller amount of the titrant as compared to if we use phenolphthalein as the indicator?

A: Absolutely brilliant! If we use phenolphthalein as the indicator for the weak base–strong acid titration, we would have "over-added" the titrant as compared to the use of methyl orange. The following tables show the appropriate indicators for different types of titration and their color changes in solutions of different pH values.

Type of titration	Marked pH change	Suitable indicator
Strong acid–strong base	4–10	Phenolphthalein, methyl orange; bromothymol blue; screened methyl orange
Strong acid–weak base	3.5–6.5	Methyl orange; screened methyl orange
Weak acid–strong base	7.5–10.5	Phenolphthalein; bromothymol blue
Weak acid–weak base	No marked change	No suitable indicator

Indicator	Approximate pH range	Color in acid solution	Color in alkaline solution
Methyl orange	3.1–4.4	Red	Yellow
Screened methyl orange	3.1–4.4	Violet	Green
Methyl red	4.2–6.3	Red	Yellow
Litmus	6.0–8.0	Red	Blue
Bromothymol blue	6.0–7.6	Yellow	Blue
Phenolphthalein	8.3–10.0	Colorless	Pink

(f) pOH can be used to measure the concentration of OH^- ions in a solution and is given by the following formula: $pOH = -\log[OH^-]$. The relationship between pH and pOH is given by: $pH + pOH = 14$.

   Calculate the pOH of a solution of sodium hydroxide of concentration $0.10\,mol\,dm^{-3}$.

Explanation:

Since NaOH is a strong base, $[OH^-] = [NaOH] = 0.10\,mol\,dm^{-3}$.

Therefore, $pOH = -\log 0.10 = 1$.

4. (a) Compound **A** is a monoacidic base. 8.63 g of compound **A** was dissolved in $250\,cm^3$ of water in a volumetric flask. $25.0\,cm^3$ of this solution required $23.15\,cm^3$ of $0.400\,mol\,dm^{-3}$ hydrochloric acid for a complete neutralization. Determine the molar mass of compound **A**.

Explanation:

A monoacidic base means one mole of the base would react with one mole of H^+ ions.

Amount of HCl used $= \dfrac{23.15}{1000} \times 0.400 = 9.26 \times 10^{-3}\,\text{mol}$.

Amount of compound **A** in 8.63 g $= \dfrac{8.63}{M}$ mol, where M = molar mass of compound **A**.

Amount of HCl used in 23.15 cm^3 = Amount of compound **A** in 25.0 cm^3.

Amount of compound **A** in 250 cm^3

$\qquad = 10 \times 9.26 \times 10^{-3}$

$\qquad = 9.26 \times 10^{-2}$ mol

$\qquad =$ Amount of compound **A** in 8.63 g

$\Rightarrow \dfrac{8.63}{M} = 9.26 \times 10^{-2}$

$\Rightarrow M = 93.2\,\text{g mol}^{-1}$.

Do you know?

— A base is the chemical opposite of an acid. Why? An acid increases the concentration of H^+ but a base *decreases* it. A reaction between an acid and a base is termed *neutralization*, producing a salt and water. Thus, metal oxides and hydroxides are considered as bases because of the reaction of the O^{2-} ion (of the oxide) and the OH^- (of the hydroxide) with the H^+ from the acid.

— A base can be soluble or insoluble; soluble bases are called *alkalis* and they dissolve in water to produce the hydroxide ion, OH^- ion:

$$NaOH(s) \rightarrow Na^+(aq) + OH^-(aq).$$

— Aqueous ammonia is a weak base because of the partial hydrolysis of ammonia in water as shown:

$$NH_3(g) + H_2O(l) \rightleftharpoons OH^-(aq) + NH_4^+(aq).$$

(Continued)

(*Continued*)

— Calcium(II) hydroxide ($Ca(OH)_2$) or slaked lime is a weak alkali in water. Due to the sparing solubility of calcium(II) hydroxide in water, it forms a dilute solution called limewater. Limewater is thus $Ca(OH)_2(aq)$. This is an important example to show that the solubility of an oxide or hydroxide in water would affect its basicity!

Q In the previous parts, we learned that the conjugate base (CH_3COO^-) of a weak acid (CH_3COOH) is basic because of its high affinity for a H^+. So, does it mean that the conjugate acid (NH_4^+) of a weak base (NH_3) is acidic because the conjugate acid has low affinity for a H^+?

A: You are absolutely right again here! A weak base is weak because the base has low affinity to accept a H^+ as compared to a strong base. This would mean that even after the weak base is protonated, it would not "hold on" to the proton long enough before it lose it again. This thus accounts for the reversible reaction!

(b) A dibasic acid has the formula H_2AO_4. A solution of the acid contains $10.0\,g\,dm^{-3}$ of H_2AO_4. In a titration, $25.0\,cm^3$ of the acid reacted with $25.30\,cm^3$ of $0.20\,mol\,dm^{-3}$ of sodium hydroxide.

(i) Calculate the concentration of the acid in $mol\,dm^{-3}$.

Explanation:

$$H_2AO_4(aq) + 2OH^-(aq) \rightarrow AO_4^{2-}(aq) + 2H_2O(l).$$

Amount of NaOH used $= \dfrac{25.30}{1000} \times 0.20 = 5.06 \times 10^{-3}\,mol.$

Amount of H_2AO_4 used $= \dfrac{1}{2} \times$ Amount of NaOH used $= 2.53 \times 10^{-3}\,mol.$

Concentration of $H_2AO_4 = 2.53 \times 10^{-3} \div \dfrac{25.0}{1000} = 0.101\,mol\,dm^{-3}.$

(ii) Calculate the relative molecular mass of the acid.

Explanation:

Let the molar mass of the acid be $M\,\mathrm{g\,mol^{-1}}$.

Concentration of $H_2AO_4 = \dfrac{10}{M} = 0.101\,\mathrm{mol\,dm^{-3}}$.

Therefore, $M = \dfrac{10}{0.101} = 98.8\,\mathrm{g\,mol^{-1}}$.

Hence, the relative molecular mass of the acid is 98.8.

Do you know?

— You can make use of a simple titration experiment to determine the molar mass or relative molecular mass of an acidic or basic compound.

(iii) Suggest an identity for the element **A**.

Explanation:

Let the molar mass of **A** be $x\,\mathrm{g\,mol^{-1}}$.

Molar mass of $H_2AO_4 = 2(1.0) + x + 4(16.0) = 98.8\,\mathrm{g\,mol^{-1}}$.

Therefore, $x = 32.8\,\mathrm{g\,mol^{-1}}$.

From the Periodic Table, **A** is probably sulfur (S).

(iv) Give the electronic configuration of element **A**.

Explanation:

Sulfur atom has 16 electrons; the electronic configuration is 2.8.6.

(v) Draw the dot-and-cross diagram for H_2AO_4.

Explanation:

Q   Why are there 12 electrons around the sulfur atom in the above dot-and-cross diagram? Why does the sulfur atom not obey the octet rule?

A: Sulfur and other Period 3 elements and beyond can have more than eight electrons in their valence electronic shell. This is because the valence shell of Period 3 elements is the $n = 3$ electronic shell, which can hold a maximum of 18 electrons $(2n^2 = 2(3)^2 = 18)$. Thus, when Period 3 elements form compounds, it does not necessary need to fulfill the octet configuration. Take note that the octet rule is not a "dictating rule" that all elements must follow!

(c) A solution is made by dissolving 7.5 g of sodium hydroxide, containing an inert impurity, in water and making up to 250 cm^3 of solution. If 20 cm^3 of this solution is exactly neutralized by 13 cm^3 of 1.0 mol dm^{-3} HNO_3, calculate the percentage purity of the sodium hydroxide.

Explanation:

$$HNO_3(aq) + NaOH(aq) \rightarrow NaNO_3(aq) + H_2O(l).$$

Amount of HNO_3 used $= \dfrac{13}{1000} \times 1.0 = 1.3 \times 10^{-2}\,mol.$

Amount of NaOH in $20\,cm^3$ = Amount of HNO_3 used $= 1.3 \times 10^{-2}\,mol.$

Amount of NaOH in $250\,cm^3 = \dfrac{250}{20} \times 1.3 \times 10^{-2} = 0.163\,mol.$

Mass of NaOH in $7.5\,g$ of impure NaOH $= 0.163 \times (23.0 + 16.0 + 1.0) = 6.5\,g.$

Percentage purity of the sodium hydroxide $= \dfrac{6.5}{7.5} \times 100 = 86.7\%.$

Do you know?

— You can make use of a simple titration experiment to determine the purity of an acidic or basic compound.

(d) Limestone is actually made up of calcium carbonate which is insoluble in water. $2.0\,g$ of the impure calcium carbonate was added to $50.0\,cm^3$ hydrochloric acid solution with a concentration of $1.0\,mol\,dm^{-3}$. The mixture was then transferred to a $250\,cm^3$ volumetric flask and topped up to the mark. $25.0\,cm^3$ of this solution was found to require $20.0\,cm^3$ of sodium hydroxide of concentration $0.20\,mol\,dm^{-3}$ for complete neutralization. Determine the percentage purity of the calcium carbonate.

Explanation:

$$HCl(aq) + NaOH(aq) \rightarrow NaCl(aq) + H_2O(l).$$

Amount of NaOH used $= \dfrac{20}{1000} \times 0.20 = 4.0 \times 10^{-3}$ mol.

Amount of HCl in $25.0\,cm^3$ = Amount of NaOH used $= 4.0 \times 10^{-3}$ mol.

Amount of leftover HCl in $250\,cm^3 = \dfrac{250}{25.0} \times 4.0 \times 10^{-3} = 4.0 \times 10^{-2}$ mol.

Initial amount of HCl used in $50.0\,cm^3 = \dfrac{50.0}{1000} \times 1.0 = 5.0 \times 10^{-2}$ mol.

Amount of HCl used to react with calcium carbonate $= 5.0 \times 10^{-2} - 4.0 \times 10^{-2} = 1.0 \times 10^{-2}$ mol.

$$2HCl(aq) + CaCO_3(s) \rightarrow CaCl_2(aq) + CO_2(g) + H_2O(l).$$

Amount of $CaCO_3$ present $= \dfrac{1}{2} \times$ Amount of HCl used to react with calcium carbonate $= \dfrac{1}{2} \times 1.0 \times 10^{-2} = 5.0 \times 10^{-3}$ mol.

Mass of $CaCO_3$ in $2.0\,g$ of impure $CaCO_3 = 5.0 \times 10^{-3} \times \{40.1 + 12.0 + 3(16.0)\} = 0.50\,g.$

Percentage purity of the $CaCO_3 = \dfrac{0.50}{2.0} \times 100 = 25.0\%.$

Do you know?

— The above titration is known as a back-titration experiment. Essentially, a compound is added to an excess but known amount of reactant. The leftover reactant is then determined through titration to find out how much is left. The actual amount of reactant that has reacted with the compound can then be back calculated.

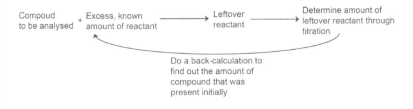

5. $25.0\,cm^3$ of $0.100\,mol\,dm^{-3}$ hydrochloric acid in a conical flask, was titrated against $0.100\,mol\,dm^{-3}$ sodium hydroxide from a burette. This is an example of a strong base–strong acid titration. The reaction can be represented by the following equation:

$$HCl(aq) + NaOH(aq) \rightarrow NaCl(aq) + H_2O(l).$$

   (a) Explain the terms *strong acid* and *strong base*.

Explanation:

A strong acid or a strong base fully dissociates in water. An example of a strong acid and a strong base are given below:

$$HCl(aq) + H_2O(l) \rightarrow H_3O^+(aq) + Cl^-(aq); \quad \text{and}$$

$$NaOH(s) + aq \rightarrow Na^+(aq) + OH^-(aq).$$

   (b) Calculate the pH of the acid solution before the titration.

Explanation:

Since HCl is a strong acid, $[HCl] = [H_3O^+] = 0.100\,mol\,dm^{-3}$.

pH of the acid solution before the titration $= -\log [H_3O^+] = -\log 0.100 = 1.0$.

   (c) When $5.0\,cm^3$ of sodium hydroxide solution was added to the acid, calculate
   (i) the number of moles of H^+ remaining in the conical flask;

Explanation:

$$HCl(aq) + NaOH(aq) \rightarrow H_2O(l) + NaCl(aq).$$

Amount of NaOH added $= \dfrac{5.0}{1000} \times 0.100 = 5.0 \times 10^{-4}\,mol.$

Amount of HCl reacted with the NaOH = Amount of NaOH added $= 5.0 \times 10^{-4}\,mol.$

Amount of HCl present initially $= \dfrac{25.0}{1000} \times 0.100 = 2.5 \times 10^{-3}\,mol.$

Amount of H^+ remaining in the conical flask $= 2.5 \times 10^{-3} - 5.0 \times 10^{-4} = 2.0 \times 10^{-3}\,mol.$

(ii) the new concentration of H^+ in the conical flask; and

Explanation:

New volume of solution $= 25.0 + 5.0 = 30.0\,cm^3.$

New concentration of H^+ in the conical flask $= 2.0 \times 10^{-3} \div \dfrac{30.0}{1000} = 6.67 \times 10^{-2}\,mol\,dm^{-3}.$

(iii) the new pH of the mixture in the conical flask.

Explanation:

The new pH of the mixture in the conical flask $= -\log[H_3O^+] = -\log 6.67 \times 10^{-2} = 1.18.$

(d) Calculate the pH of the mixture after $24.0\,cm^3$ of $0.100\,mol\,dm^{-3}$ sodium hydroxide was added.

Explanation:

Amount of NaOH added = $\dfrac{24.0}{1000} \times 0.100 = 2.4 \times 10^{-3}$ mol.

Amount of HCl reacted with the NaOH = Amount of NaOH added = 2.4×10^{-3} mol.

Amount of HCl present initially = $\dfrac{25.0}{1000} \times 0.100 = 2.5 \times 10^{-3}$ mol.

Amount of H^+ remaining in the conical flask = $2.5 \times 10^{-3} - 2.4 \times 10^{-3} = 1.0 \times 10^{-4}$ mol.

The new concentration of H^+ in the conical flask = $1.0 \times 10^{-4} \div \dfrac{49.0}{1000} = 2.04 \times 10^{-3}$ mol dm^{-3}.

The new pH of the mixture in the conical flask = $-\log[H_3O^+]$ = $-\log 2.04 \times 10^{-3} = 2.69$.

(e) Would the pH of the mixture in the conical flask after 26.0 cm^3 of sodium hydroxide has been added be closer to pH 7 or 11? Explain your choice.

Explanation:

Amount of NaOH added = $\dfrac{26.0}{1000} \times 0.100 = 2.6 \times 10^{-3}$ mol.

Amount of HCl reacted with the NaOH = Amount of NaOH added = 2.5×10^{-3} mol.

Amount of NaOH left = $2.6 \times 10^{-3} - 2.5 \times 10^{-3} = 1.0 \times 10^{-4}$ mol.

Since there is an excess NaOH that is unreacted, the pH of the solution would be closer to 11 and not neutral at pH 7.

Q   How do you calculate the pH of the above alkaline solution?

A: The new concentration of OH^- in the conical flask $= 1.0 \times 10^{-4} \div \dfrac{51.0}{1000}$

$$= 1.96 \times 10^{-3} \, mol \, dm^{-3}.$$

pOH $= -\log [OH^-] = -\log 1.96 \times 10^{-3} = 2.71$.

Since pH + pOH = 14, therefore pH = 14 − 2.71 = 11.29.

(f) Outline the most appropriate method to prepare a pure dry sample of sodium chloride starting with dilute hydrochloric acid.

Explanation:

To prepare a pure dry sample of sodium chloride starting with dilute hydrochloric acid:

Step 1: To get the salt solution, react an equimolar amount of HCl(aq) and NaOH(aq).

Step 2: Concentrate the solution through evaporation.

Step 3: Crystallize out the salt through cooling the hot solution.

Step 4: Filter and you get the salt crystals!

6. Calcium dihydrogen phosphate, $Ca(H_2PO_4)_2$, is commonly found in fertilizers.

   (a) Explain why calcium dihydrogen phosphate is both an *acid* and a *salt*.

Explanation:

Calcium dihydrogen phosphate is considered a salt as it is an *ionic* compound that is formed when phosphoric(V) acid (H_3PO_4) reacts with a base or a metal. For example,

$$Ca(OH)_2 + 2H_3PO_4 \rightarrow Ca(H_2PO_4)_2 + 2H_2O; \quad \text{and}$$

$$Ca + 2H_3PO_4 \rightarrow Ca(H_2PO_4)_2 + H_2.$$

Calcium dihydrogen phosphate is considered an acid as it can dissolve in water to give an acidic solution due to the following dissociation:

$$H_2PO_4^-(aq) + H_2O(l) \rightleftharpoons HPO_4^{2-}(aq) + H_3O^+(aq).$$

Q

So, does that mean that phosphoric(V) acid can in fact undergo the following three dissociations?

$$H_3PO_4(aq) + H_2O(l) \rightarrow H_2PO_4^-(aq) + H_3O^+(aq)$$

$$H_2PO_4^-(aq) + H_2O(l) \rightarrow HPO_4^{2-}(aq) + H_3O^+(aq)$$

$$HPO_4^{2-}(aq) + H_2O(l) \rightarrow PO_4^{3-}(aq) + H_3O^+(aq).$$

Does each of the above three dissociations occur to the same extent?

A: Phosphoric(V) acid is a tribasic acid which can dissociate in three stages as shown above. But the amount of dissociation for each stage is not to the same extent. It is logical to understand why: imagine after the first dissociation, for example $H_3PO_4 \rightarrow H^+ + H_2PO_4^-$, you have the negatively charged $H_2PO_4^-$ ion formed. Do you think it is as easy to remove another H^+ ion from a negatively charged species as compared to from a neutral H_3PO_4? Of course not! Hence, the second or third dissociation usually occurs to a lesser extent than the previous dissociation.

(b) Calcium dihydrogen phosphate can be made by reacting calcium hydroxide with phosphoric(V) acid, H_3PO_4, a tribasic acid.

(i) Write an equation for the reaction of calcium hydroxide with phosphoric acid.

Explanation:

$$Ca(OH)_2 + 2H_3PO_4 \rightarrow Ca(H_2PO_4)_2 + 2H_2O.$$

(ii) Give the formulae of two other calcium salts formed from calcium hydroxide and phosphoric(V) acid.

Explanation:

$$Ca(OH)_2 + H_3PO_4 \rightarrow CaHPO_4 + 2H_2O; \quad \text{and}$$

$$3Ca(OH)_2 + 2H_3PO_4 \rightarrow Ca_3(PO_4)_2 + 6H_2O.$$

(c) The pH of a $0.25 \, mol \, dm^{-3}$ solution of calcium dihydrogen phosphate has a value of 4.4, while that of a $0.50 \, mol \, dm^{-3}$ solution of ethanoic acid has a pH of 3.8.

(i) Explain which acid, calcium dihydrogen phosphate or ethanoic acid, is a weaker acid.

Explanation:

A lower pH value would mean a higher concentration of H^+ ions.

If the calcium dihydrogen phosphate in the $0.25 \, mol \, dm^{-3}$ solution has fully dissociated, then $[H^+] = 0.50 \, mol \, dm^{-3}$. Hence, the pH should be 0.30.

Similarly, if the ethanoic acid in the $0.50\,\text{mol}\,\text{dm}^{-3}$ solution has fully dissociated, then $[\text{H}^+] = 0.50\,\text{mol}\,\text{dm}^{-3}$. Hence, the pH should be 0.30.

But since the pH value of ethanoic acid (3.8) is lower than that of the calcium dihydrogen phosphate (4.4), we could probably say that calcium dihydrogen phosphate is a weaker acid.

 What do you mean when you say that "we could probably say that calcium dihydrogen phosphate is a weaker acid"?

A: We can only use pH value to gauge the strength of an acid provided that you are comparing two acids of the *same* concentrations. Only then would it be fair to conclude that the stronger acid would be the one that dissociates more and thus gives a higher $[\text{H}_3\text{O}^+]$.

 But how does the concentration of the weak acid affects the dissociation and hence the concentration of the H^+ ions?

A: Assuming that we have the following weak acid, HA, dissociating in water:

$$\text{HA(aq)} + \text{H}_2\text{O(l)} \rightleftharpoons \text{H}_3\text{O}^+\text{(aq)} + \text{A}^-\text{(aq)}.$$

The amount of the weak acid dissociated depends very much on its concentration. How? According to Le Chatelier's Principle, a higher concentration of a weak acid would cause it to dissociate to a larger extent, hence the amount of dissociation increases, which leads to an increase in $[\text{H}_3\text{O}^+]$ and decrease in pH. But there is a limit to the increase in the amount of a weak acid that has dissociated and a decrease in pH! This is because as a larger amount of a weak acid dissociates, more free ions are formed. It would also mean that the backward reaction is also more likely to occur. From the above equation, we can see very clearly that in order for HA to function as an acid, we need a base (H_2O) to be present. Imagine if there are too many molecules of a weak acid but not enough H_2O, can the weak acid dissociate? So, we can only use pH to compare the strengths of two weak acids provided they are of equal concentrations. Secondly, both must be of a similar type; for example, both must be monobasic, i.e., one mole of the acid reacts with one mole of OH^-.

Q A higher concentration of a weak acid causes more weak acid to dissociate and at the same time, causes the rate of the backward reaction to increase. Then, at a lower concentration value, when the weak acid dissociates, there is less chance for the backward reaction to occur because the volume of the system is larger. So, would this cause more $H_3O^+(aq)$ to form and hence lower the pH value?

A: There is a fallacy here! Yes, at a lower concentration of the weak acid, you can perceive that dissociation somehow increases because the rate of the backward reaction decreases, so the number of moles of $H_3O^+(aq)$ increases. But do not forget that the volume of the system also increases because of dilution. Thus, when you calculate the pH, which is pH = $-\log[H_3O^+]$ = $-\log\left(\dfrac{\text{number of moles } H_3O^+}{\text{volume of solution}}\right)$, the increase in the volume is much more signification than the number of moles. Hence, overall, the pH increases. Just imagine, if the solution of the weak acid is greatly diluted, just like in pure water, would the pH just be about 7 instead?

(ii) The pH of $1.0\,\text{mol dm}^{-3}$ solution of sulfuric acid is 0.30. Rank in order of increasing acidity for sulfuric acid, calcium dihydrogen phosphate, and ethanoic acid.

Explanation:

A lower pH value would mean a higher concentration of H^+ ions.

If sulfuric acid in the $1.0\,\text{mol dm}^{-3}$ solution has fully dissociated, then $[H^+] = 2.0\,\text{mol dm}^{-3}$. Hence, the pH should be 0.30, which coincides with the measurement. Thus, sulfuric acid is a strong acid!

Therefore, based on the given information, in order of increasing acidity: calcium dihydrogen phosphate < ethanoic acid < sulfuric acid.

(iii) Describe an experiment, other than measuring pH, which you would carry out to show that sulfuric acid is a strong acid but ethanoic acid is a weak acid. Show what measurements you would make and what results you would expect.

Explanation:

The idea that we are using to determine the strength of the acid is simple: the stronger the acid, the greater the amount of dissociated H^+. Thus, if we put in a fixed amount of solid $CaCO_3$ into same volume of each of the two acids, we can measure the volume of CO_2 gas that is collected with respect to time. The stronger the acid, the faster the rate of production of the CO_2.

But, we have to take note that sulfuric acid is a dibasic acid while ethanoic acid is a monobasic acid. Hence, the concentration of sulfuric acid must be half that of ethanoic acid.

 Q Other than the above method, is there any other way to differentiate the strengths of the two acids?

A: Well, you can add a known volume of sodium hydroxide solution into the same volume of each of the two acids. Then, measure the temperature change for the acid–base reaction. Since the strong acid has fully dissociated in water, the energy change for the acid–base reaction would be more exothermic than that between the weaker acid and the base. This is because since the weak acid has not fully dissociated in water, part of the heat energy that is released during the acid–base reaction would be diverted to "help" dissociate the undissociated weak acid molecules. Thus, the temperature change for the acid–base reaction between the weaker acid and the base would be smaller than that between the strong acid and the base.

7. Describe how you would prepare crystals of each of the following salt using the method indicated. Write an equation for each reaction.

   (a) Anhydrous iron(III) chloride, $FeCl_3$, by direct reaction of the elements.

Explanation:

Dry chlorine gas can be passed over heated iron.

$$2Fe(s) + 3Cl_2(g) \rightarrow 2FeCl_3(s).$$

The mixture is then dissolved in water to get rid of the unreacted iron metal. The solution is filtered, then concentrated through evaporation. Next, crystallize out the salt through cooling the hot solution. Filter and you get the salt crystals!

(b) Sodium nitrate, $NaNO_3$, by reacting an acid with an alkali.

Explanation:

Step 1: To get the salt solution, react an equimolar amount of $HNO_3(aq)$ and $NaOH(aq)$.

$$HNO_3(aq) + NaOH(aq) \rightarrow NaNO_3(aq) + H_2O(l).$$

Step 2: Concentrate the solution through evaporation.

Step 3: Crystallize out the salt through cooling the hot solution.

Step 4: Filter and you get the salt crystals!

(c) Copper(II) sulfate, $CuSO_4 \cdot 5H_2O$, by reacting an acid with a metal oxide.

Explanation:

Step 1: To get the salt solution, react excess $CuO(s)$ with sulfuric acid.

$$CuO(s) + H_2SO_4(aq)$$
$$\rightarrow CuSO_4(aq) + H_2O(l) + \text{unreacted oxide}.$$

Step 2: Filter the mixture and collect the filtrate.

Step 3: Concentrate the filtrate through evaporation.

Step 4: Crystallize out the salt through cooling the hot solution.

Step 5: Filter and you get the salt crystals!

 Why is excess metal oxide used when preparing the salt?

A: As the metal oxide used is insoluble, when we used an excess of these substances in the preparation of the salt, we ensure that all the acid used is completely reacted. If not, during the crystallization of the salt, it might be contaminated by the leftover acid.

Do you know?

— When there are water molecules embedded within the giant ionic lattice structure of a salt, we have the formation of a *hydrated salt*. These water molecules are known as the *water of crystallization*. The water molecules are attracted to the ions of the ionic lattice structure via *ion–dipole* interactions as the water molecules have *permanent* dipole (refer to Chapter 3 on Chemical Bonding). But nevertheless, the water of crystallization can be easily removed via heating. For example:

$$CuSO_4 \cdot 5H_2O(s) \rightarrow \qquad CuSO_4(s) \qquad + 5H_2O(l)$$

   Blue hydrated form   White anhydrous form

— The desiccant, silica gel, has cobalt(II) chloride embedded within the giant *covalent* lattice structure of the SiO_2. The hydrated form of cobalt(II) chloride is $CoCl_2 \cdot 6H_2O$ which is *pink* in color. When $CoCl_2 \cdot 6H_2O$ loses its water of crystallization, it reverts back to the *blue* anhydrous $CoCl_2$. Thus, the color of the cobalt(II) chloride serves as an indicator as to when the drying agent needs to be changed.

CHAPTER 9

REDOX REACTIONS

1. Hydrogen reacts explosively with fluorine in the dark to produce a gas, which on dissolving in water gives hydrofluoric acid:

$$H_2(g) + F_2(g) \rightarrow 2HF(g) + aq \rightarrow 2HF(aq).$$

(a) Give the oxidation states of hydrogen and fluorine in hydrogen fluoride.

Explanation:

The oxidation states of hydrogen and fluorine in hydrogen fluoride are +1 and −1, respectively.

Q What is an *oxidation state*?

A: An oxidation state indicates the *ease* of the atom in a species (which can be an atom, ion, or molecule) to undergo *oxidation*. A *positive* oxidation state indicates that the atom would be oxidized, whereas a *negative* oxidation state indicates reduction. The oxidation state is indicated by the *oxidation number*.

Q So, if an atom has an oxidation number of +1, does it mean that it has lost an electron?

A: No, that is not necessarily true. The oxidation number of Na^+ is +1 and indeed the Na atom has lost an electron to become Na^+. As for the covalent molecule HF, the H atom has also an oxidation number of +1, yet it has not completely lost any electrons at all. Similarly, the oxidation number of F in HF is −1. This does not mean that the F atom has gained an electron.

So, remember that the oxidation number indicates the *ease* of the atom to lose electrons, but it does not necessarily mean that the atom has already lost electrons. Thus, *the oxidation number may not be equivalent to the formal charge* that an atom carries.

 Q So, how does one determine the oxidation number of an atom in a species?

A: If it is a cation or anion, then the oxidation number would be equivalent to the formal charge, such as Na^+ and Cl^-.

If the atom is covalently bonded to other atoms, one needs to analyze each of the covalent bonds in turn, and determine which of the two bonding atoms has a higher electronegativity value. The one that is *more* electronegative would *polarize* the shared electron cloud more toward itself. Hence, this atom would have a *negative* oxidation number. This would mean that it has a potential to undergo reduction as compared to the other less electronegative atom. The oxidation number would then be determined by the number of electrons contributed by this atom in the bond-sharing process.

Do you know?

— To determine the oxidation state:

(1) The oxidation state of atoms in the *elemental form* is being assigned a zero value. E.g., Ca, F atoms in Ca(s), and $F_2(g)$.

(2) In all compounds, fluorine has the oxidation state of −1.

(3) In all compounds, hydrogen has the oxidation state of +1. An exception will be in the case of metal hydrides such as NaH, where the oxidation state of H is −1.

(4) In all compounds, oxygen has the oxidation state of −2. An exception will be in the case of peroxides such as H_2O_2, where the oxidation state of O is −1 and in OF_2 where it is +2.

(Continued)

(*Continued*)

(5) For a pair of covalently bonded atoms with a single bond, the more electronegative atom is assigned an oxidation state of −1 and the less electronegative atom has an oxidation state of +1.

> Recall that the electronegativity of elements
> - increases across the period, and
> - decreases down the group.

(6) The sum of the oxidation states of all atoms in a neutral compound (e.g., KCl and CO_2) is zero.
(7) For a monoatomic ion, its oxidation state corresponds to the net charge on the ion.
(8) For a polyatomic ion, i.e., an ion with more than three atoms, the sum of the oxidation states of all atoms corresponds to the net charge on the ion.

 Q So, if two dissimilar atoms form a double bond, the more electronegative atom would have an oxidation state of −2?

A: You are right! The less electronegative one would have an oxidation number of +2. Likewise, for a triple bond, the more electronegative atom would have an oxidation number of −3.

(b) Identify both the oxidizing agent and the reducing agent.

Explanation:

An oxidizing agent itself undergoes reduction, while a reducing agent itself undergoes oxidation. Now, since hydrogen undergoes oxidation, it is acting as a reducing agent. Likewise, fluorine is the oxidizing agent in the above reaction.

Do you know?

— Specifically, a redox reaction involves *electron transfer* between a pair of substances. In the process of electron transfer, one species will *lose* electron(s) to another species, which in turn will *gain* the electron(s).

  • The species that loses electron(s) is said to be oxidized, i.e., it has undergone oxidation. This species is known as the *reducing agent* or a *reductant*.
  • The species that accepts the electron(s) is said to be reduced, i.e., it has undergone reduction. This species is known as the *oxidizing agent* or *oxidant*.

— In a nutshell, an oxidizing agent will *itself* undergo reduction. A reducing agent will *itself* be oxidized. And in a redox reaction, both the reducing agent and oxidizing agent must be present together! Otherwise who would provide the electrons and who would take up the electrons? So, do you see that this is very similar to an acid–base reaction? An acid must be present in order for the base to behave as a base, and vice versa!

(c) Explain, in terms of electron gain and loss, why this is a redox reaction.

Explanation:

The oxidation number of hydrogen increases from 0 to +1, which indicates that hydrogen has "lost" an electron. In contrast, the oxidation number of fluorine has decreased from 0 to −1, an indication of a "gain" of an electron. Thus, the above reaction is a redox reaction.

(d) Give the dot-and-cross diagram for hydrogen fluoride.

Explanation:

$$H {\scriptstyle\times}\ddot{\underset{..}{F}}:$$

> (e) Explain why hydrogen fluoride is a liquid at room temperature.

Explanation:

Hydrogen fluoride is a liquid at room temperature because of the strong hydrogen bonds between the HF molecules.

$$\overset{\delta+}{H}{-}\overset{\delta-}{\ddot{F}}: \cdots\cdots \overset{\delta+}{H}{-}\overset{\delta-}{F}$$

hydrogen bond

> (f) Given the following information,
>
> | Covalent bond | Bond energy in kJ mol^{-1} |
> |:---:|:---:|
> | H–H | 436 |
> | F–F | 151 |
> | H–F | 299 |
>
> calculate ΔH for the reaction between hydrogen and fluorine to give gaseous hydrogen fluoride.

Explanation:

$$H{-}H + F{-}F \rightarrow 2H{-}F.$$

Total energy absorbed during bond breaking = BE(H–H) + BE(F–F) = $436 + 151 = 587$ kJ mol^{-1}.

Total energy released during bond forming = $2 \times$ BE(H–F) = 598 kJ mol^{-1}.

$$\Delta H = \text{(Sum of total energy for bond breaking)}$$
$$- \text{(Sum of total energy for bond making)}.$$

Hence, $\Delta H = 587 - 598 = -11$ kJ mol^{-1}.

(g) Draw an energy profile diagram for the formation of hydrogen fluoride gas. Indicate on the diagram the activation energy and the ΔH for the reaction.

Explanation:

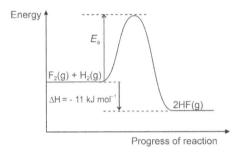

2. Iron(II) sulfate is an important component in moss killers. The percentage by mass of iron in a sample of moss killer can be determined by titration against acidified potassium manganate(VII).

   (a) The ionic equation for the redox reaction between acidified MnO_4^- and Fe^{2+} is given below:

$$MnO_4^-(aq) + 8H^+(aq) + 5Fe^{2+}(aq)$$
$$\rightarrow Mn^{2+}(aq) + 4H_2O(l) + 5Fe^{3+}(aq).$$

   Explain, in terms of changes in oxidation number, why this reaction involves both oxidation and reduction. Hence, identify both the oxidizing agent and reducing agent.

Explanation:

The oxidation number of Mn in MnO_4^-(aq) decreases from a +7 to +2 in Mn^{2+}. This is a reduction process; hence, MnO_4^-(aq) is an oxidizing agent.

The oxidation number of Fe^{2+}(aq) increases from a +2 to +3 in Fe^{3+}. This is an oxidation process; hence, Fe^{2+}(aq) is a reducing agent.

Do you know?

— We can represent the sub-processes of a redox equation through *half-equations*. For example, the following is the ionic equation for the reaction between zinc metal and copper(II) sulfate ($CuSO_4$) solution:

$$Zn(s) + Cu^{2+}(aq) \rightarrow Zn^{2+}(aq) + Cu(s).$$

To represent the oxidation of Zn, we have the oxidation half-equation:

$$Zn(s) \rightarrow Zn^{2+}(aq) + 2e^-,$$

where e^- represents a negatively charged electron. Do you see that in an oxidation half-equation, the electrons are on the *right*-hand side of the equation? This is an indication of the *release* of electrons during an oxidation reaction.

To represent the reduction of Cu^{2+}, we have the reduction half-equation:

$$Cu^{2+}(aq) + 2e^- \rightarrow Cu(s).$$

Do you see that in a reduction half-equation, the electrons are on the *left*-hand side of the equation? This is an indication of the "*consumption*" of electrons during a reduction reaction. If we combine these two half-equations, we get the overall redox equation! Take note that in a balanced redox equation, the amount of electrons released during oxidation MUST be all taken in for reduction. There cannot be "unconsumed" electrons floating around!

(b) A 9.50 g sample of the moss killer was dissolved in water to make a 20.00 cm^3 solution. This solution requires 22.50 cm^3 of 0.750 mol dm^{-3} MnO_4^- to reach the endpoint.

   (i) Give the color change for MnO_4^-(aq) and explain why an indicator is not needed for the above titration.

Explanation:

In an acidic medium, the manganate(VII) ion is reduced as follows:

$$MnO_4^-(aq) + 8H^+(aq) + 5e^- \rightarrow Mn^{2+}(aq) + 4H_2O(l).$$

purple   faint pink

Since $KMnO_4(aq)$ is purple in color while its reduced product, $Mn^{2+}(aq)$, in an acidified medium is *essentially* colorless, an external indicator is not required to be added in the course of the titration. The endpoint is taken when the first permanent pink coloration is formed because of an excess drop of potassium manganate(VII) is added.

 In the above equation, $Mn^{2+}(aq)$ is said to have a faint pink coloration. So, shouldn't the formation of the $Mn^{2+}(aq)$ be used as an indication for the endpoint of titration?

A: In reality, the faint pink coloration of $Mn^{2+}(aq)$ is virtually undetectable.

(ii) Calculate the amount of MnO_4^- used in the titration.

Explanation:

Amount of MnO_4^- used in the titration $= \dfrac{22.5}{1000} \times 0.75 = 1.69 \times 10^{-2}$ mol.

(iii) Hence, calculate the mass of iron present in the 9.50 g sample of moss killer.

Explanation:

$$MnO_4^-(aq) + 8H^+(aq) + 5Fe^{2+}(aq) \rightarrow Mn^{2+}(aq) + 4H_2O(l) + 5Fe^{3+}(aq).$$

Amount of Fe^{2+} present $= 5 \times$ Amount of MnO_4^- used in the titration $= 5 \times 1.69 \times 10^{-2} = 8.45 \times 10^{-2}$ mol.

Mass of iron present in the 9.50 g sample of moss killer = $8.45 \times 10^{-2} \times 55.8 = 4.72$ g.

(iv) Determine the percentage by mass of iron in the moss killer.

Explanation:

Percentage by mass of iron in the moss killer = $\dfrac{4.72}{9.50} \times 100 = 49.6\%$.

(v) If the iron(II) sulfate also reacts with ammonium dichromate(VI) as follows:

$$6Fe^{2+}(aq) + Cr_2O_7^{2-}(aq) + 14H^+(aq) \rightarrow 6Fe^{3+}(aq) + 2Cr^{3+}(aq) + 7H_2O(l),$$

determine the volume of ammonium dichromate(VI) of concentration 0.750 mol dm^{-3} that is required to completely oxidize all the iron(II) sulfate.

Explanation:

Amount of Fe^{2+} present = 8.45×10^{-2} mol.

Amount of $Cr_2O_7^{2-}$ used = $\dfrac{1}{6} \times$ Amount of Fe^{2+} present = $\dfrac{1}{6} \times 8.45 \times 10^{-2} = 1.41 \times 10^{-2}$ mol.

Volume of ammonium dichromate(VI) needed = $\dfrac{0.0141}{0.75} = 0.0188$ dm^3 = 18.8 cm^3.

(vi) Ammonium dichromate(VI) is decomposed by heat:

$$(NH_4)_2Cr_2O_7(s) \rightarrow Cr_2O_3(s) + 4H_2O(l) + N_2(g).$$

Use changes in oxidation state to explain why this is a redox reaction.

Explanation:

The oxidation number of N in NH_4^+ increases from -3 to 0 in N_2. This is an oxidation process. Hence, NH_4^+ is a reducing agent.

The oxidation number of Cr in $Cr_2O_7^{2-}$ decreases from a $+6$ to $+3$ in Cr_2O_3. This is a reduction process. Hence, $Cr_2O_7^{2-}$ is an oxidizing agent.

Therefore, the above reaction is a redox reaction.

Q   Can the above reaction be termed a "disproportionation" reaction?

A: A disproportionation reaction is a redox reaction in which a single substance is simultaneously being oxidized and reduced. Since the above redox reaction does not involve the simultaneous oxidation (NH_4^+) and reduction ($Cr_2O_7^{2-}$) of the same species, it is not a disproportionation reaction.

(vii) Give the electronic configurations for Fe^{2+}, Fe^{3+}, and Mn^{2+}.

Explanation:

The electronic configurations of: Fe^{2+} is 2.8.14; Fe^{3+} is 2.8.13; and Mn^{2+} is 2.8.13.

Do you know?

— When the particles have the same number of electrons, they are termed "isoelectronic." The unique feature of isoelectronic particles is that they have the same amount of inter-electronic repulsion.

(viii) Usually, for such titrations, the solution needs to be acidified by an acid. Explain why sulfuric(VI) acid is commonly used for acidification.

Explanation:

Sulfuric(VI) acid itself will not act as an oxidizing or reducing agent during the titration process.

 Q Can we use other acids such as HCl(aq) or HNO_3(aq)?

A: No! MnO_4^- is strong enough to oxidize the Cl^- ion to the chlorine gas (Cl_2). So, if you acidify it using HCl(aq), the oxidizing agent would not be effective anymore. This would also mean that we can use acidified potassium manganate(VII) to test for HCl gas! As for HNO_3, it can also act as an oxidizing agent due to the presence of the NO_3^- ion.

---

3. Aqueous iron(III) ions and aqueous tin(II) ions undergo a redox reaction according to the equation:

$$2Fe^{3+}(aq) + Sn^{2+}(aq) \rightarrow 2Fe^{2+}(aq) + Sn^{4+}(aq).$$

(a) Give both the oxidation and reduction half-equations.

Explanation:

Oxidation half-equation: $Sn^{2+}(aq) \rightarrow Sn^{4+}(aq) + 2e^-$.
Reduction half-equation: $Fe^{3+}(aq) + e^- \rightarrow Fe^{2+}(aq)$.

Q How do we derive the half-equations for a more complex redox equation?

A: Consider the redox equation

$$MnO_4^-(aq) + 8H^+(aq) + 5Fe^{2+}(aq) \rightarrow Mn^{2+}(aq) + 4H_2O(l) + 5Fe^{3+}(aq),$$

we can derive the half-equations as follows:

Steps involved	Illustration
1. Assign oxidation states to determine which reactants undergo reduction and oxidation.	$MnO_4^-(aq) + Fe^{2+}(aq) \rightarrow Mn^{2+}(aq) + Fe^{3+}(aq)$ $+7\ -2 \qquad +2 \qquad\qquad +2 \qquad\quad +3$
2. Construct two half-equations that show the specific species that are reduced or oxidized to the corresponding products.	Reduction half-equation: $MnO_4^- \rightarrow Mn^{2+}$ Oxidation half-equation: $Fe^{2+} \rightarrow Fe^{3+}$
3. Balance each of the half-equation by following the simple rules:	
• Balance the element that undergoes oxidation or reduction first.	$MnO_4^- \rightarrow Mn^{2+}$
• Balance O atoms by adding the same number of H_2O molecules to the other side of the equation.	$MnO_4^- \rightarrow Mn^{2+}$ $MnO_4^- \rightarrow Mn^{2+} + 4H_2O$
• Balance H atoms by adding H^+ ions to the other side of the equation.	$MnO_4^- \rightarrow Mn^{2+} + 4H_2O$ $MnO_4^- + 8H^+ \rightarrow Mn^{2+} + 4H_2O$
• Finally, balance charges by adding electrons.	$\underbrace{MnO_4^- + 8H^+}_{-1+8(+1)=+7} \rightarrow \underbrace{Mn^{2+} + 4H_2O}_{+2 + 4(0) = +2}$ $\underbrace{MnO4^- + 8H^+ + 5e^-}_{-1 + 8(+1) + 5(-1) = +2} \rightarrow \underbrace{Mn^{2+} + 4H_2O}_{+2 + 4(0) = +2}$
4. Repeat Step 3 for the other half-equation.	$Fe^{2+} \rightarrow Fe^{3+} + e^-$
5. Ensure that the number of electrons for each half-equation is the same by scaling one or both of these equations by appropriate multiples. i.e., no. of e^- lost = no. of e^- gained	$MnO_4^- + 8H^+ + 5e^- \rightarrow Mn^{2+} + 4H_2O$ (1) $Fe^{2+} \rightarrow Fe^{3+} + e^-$ (2) Multiply (2) by 5, $5Fe^{2+} \rightarrow 5Fe^{3+} + 5e^-$ (3)
6. Add the two half-equations together to obtain the overall balanced equation.	Add half-equations (1) and (3), $MnO_4^- + 8H^+ + 5e^- \rightarrow Mn^{2+} + 4\ H_2O$ (1) $5Fe^{2+} \rightarrow 5Fe^{3+} + 5e^-$ (3)
• Simplify the equation by removing common terms that appear on both sides of the equation. • Double-check that the number of atoms and charges are balanced on both sides of the equation. • Reduce coefficients into simplest ratio.	Balanced redox equation: $5Fe^{2+}(aq) + MnO_4^-(aq) + 8H^+(aq) \rightarrow$ $\quad 5Fe^{3+}(aq) + Mn^{2+}(aq) + 4H_2O(l)$ *Note*: Electrons should not appear in overall equation.

(b) When aqueous potassium iodide was added to aqueous iron(III) sulfate, $Fe_2(SO_4)_3$, a brown solution of iodine was formed.

(i) Explain the formation of the iodine.

Explanation:

The iodine was formed from the oxidation of the iodide by the Fe^{3+} ion.

(ii) Give both the oxidation and reduction half-equations.

Explanation:

Oxidation half-equation: $2I^-(aq) \rightarrow I_2(aq) + 2e^-$.
Reduction half-equation: $Fe^{3+}(aq) + e^- \rightarrow Fe^{2+}(aq)$.

(iii) Hence, give the overall ionic equation.

Explanation:

Overall ionic equation: $2I^-(aq) + 2Fe^{3+}(aq) \rightarrow I_2(aq) + 2Fe^{2+}(aq)$.

Q   So, does that mean we cannot synthesize FeI_3 compound?

A: You are right! You can make $FeCl_3$ and $FeBr_3$ but you just can't make FeI_3, because the moment you put I^- and Fe^{3+} together, they undergo a redox reaction!

(iv) Draw the dot-and-cross diagrams for iodine and potassium iodide.

Explanation:

$$\left[\text{K}\right]^{+} \left[\text{:}\ddot{\underset{..}{\text{I}}}\text{:}\right]^{-} \qquad \text{:}\ddot{\underset{xx}{\text{I}}}\overset{xx}{\underset{xx}{\times\text{I}}}\times$$

> (v) Explain why iodine and potassium iodide have such a great difference in melting points.

Explanation:

Iodine is a non-polar molecule with weak intermolecular forces of the instantaneous dipole–induced dipole (id–id) type. For potassium iodide, it is an ionic compound with strong ionic bonding. Hence, potassium iodide has a much higher melting point than iodine.

> (vi) Explain why solid potassium iodide is an electrical insulator but molten potassium iodide is not.

Explanation:

In the solid state, the ions in potassium iodide are rigidly held in fixed positions in the solid lattice structure. Hence, these ions are not mobile and cannot act as charge carriers. But in the molten state, the ions are mobile enough to act as charge carriers.

> 4. (a) Sulfur dioxide is a common air pollutant emitted from vehicles and power plant that uses fossil fuels to generate energy. This pollutant can be tested by passing polluted air into an acidified solution of potassium dichromate(VI). The ionic equation for this reaction is:
>
> $$3SO_2(g) + Cr_2O_7^{2-}(aq) + 2H^+(aq)$$
> $$\rightarrow 3SO_4^{2-}(aq) + 2Cr^{3+}(aq) + H_2O(l).$$
>
> (i) What would be the expected observation for a positive test?

Explanation:

If the orange $Cr_2O_7^{2-}$ solution turns green, it indicates that $Cr_2O_7^{2-}$ has been reduced to Cr^{3+}. This means that the unknown substance has caused the reduction of the $Cr_2O_7^{2-}$ ion, and hence contains a reducing agent.

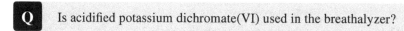

Q Is acidified potassium dichromate(VI) used in the breathalyzer?

A: Yes! Ethanol can be oxidized by potassium dichromate(VI) and the amount of Cr^{3+} formed, which is indicated by the intensity of the green coloration, tells us the amount of ethanol in the exhaled air.

(ii) What reaction has occurred to give the observation in part (a)?

Explanation:

A redox reaction has occurred to give the observation in part (a). The $Cr_2O_7^{2-}$(aq) undergoes reduction while the SO_2 undergoes oxidation.

(iii) A 3 dm^3 sample of polluted air contains 5.2% sulfur dioxide by volume (measured at s.t.p.). Calculate the volume of 0.20 mol dm^{-3} aqueous dichromate(VI), $Cr_2O_7^{2-}$, required to react with all the sulfur dioxide in the sample.

Explanation:

At s.t.p. (0°C and 1 bar), 1 mole of gas occupies a volume of 22.7 dm^3.

Thus, 3 dm^3 of gas would contain $= \dfrac{3}{22.7} \times 1 = 0.132$ mol.

Amount of SO_2 in 3 dm^3 $= \dfrac{5.2}{100} \times 0.132 = 6.87 \times 10^{-3}$ mol.

Three moles of SO_2 react with one mole of $Cr_2O_7^{2-}$.

Amount of $Cr_2O_7^{2-}$ needed = $\frac{1}{3}$ × 6.87 × 10^{-3} = 2.29 × 10^{-3} mol.

Volume of 0.20 mol dm^{-3} aqueous dichromate(VI), $Cr_2O_7^{2-}$, required = $\frac{0.00229}{0.20}$ = 0.0115 dm^3.

(b) Calcium is one of the reactive metals in Group 2.

   (i) Describe the bonding present in solid calcium and explain why calcium can conduct electricity.

Explanation:

Calcium atoms are bonded together via metallic bonds. It is the electrostatic attractive force between the positive ions and the sea of delocalized valence electrons.

    Calcium can conduct electricity because the delocalized valence electrons can act as charge carriers under the influence of an applied electrical potential difference.

   (ii) Write an equation for the reaction of calcium in water. Identify the oxidation numbers of all the atoms involved by writing the oxidation numbers underneath each symbol in the equation.

Explanation:

Oxidation number of: $Ca(s) + 2H_2O(l) \rightarrow Ca(OH)_2(aq) + H_2(g).$

                              0   +1 –2   +2 –2 +1   0

Q Is water acting as an acid in the above reaction?

A: Yes, you are right! According to the properties of an acid, water is acting as an acid because it reacts with a metal to give off hydrogen gas. But take note that water is not a very strong acid that would react with all metals. Water is a weak acid that only reacts with very reactive metals.

Q So why does pure water at 25°C not have a pH < 7? In fact, why is it pH = 7?

A: At 25°C, water molecules actually undergo *auto-ionization*:

$$H_2O(l) + H_2O(l) \rightleftharpoons H_3O^+(aq) + OH^-(aq).$$

$[H_3O^+(aq)] = [OH^-(aq)] = 10^{-7}$ mol dm^{-3} at 25°C, i.e., the rationale why the pH of pure water is 7! $[H_3O^+(aq)]$ varies depending on whether an acid or base is added to the pure water.

(iii) State which substance in part (b)(ii) has been oxidized and write a half-equation to show the oxidation process.

Explanation:

The oxidation state of calcium increases from 0 to +2; thus, calcium metal has been oxidized.

Oxidation half-equation: $Ca \rightarrow Ca^{2+} + 2e^-$.

(iv) What is the common name of the solution formed by the reaction of calcium with water?

Explanation:

The common name of the solution formed by the reaction of calcium with water is called *"limewater."*

 Q Isn't calcium hydroxide partially soluble in water?

A: Yes, calcium hydroxide is not very soluble in water but we can still dissolve some to get a decent amount of calcium hydroxide solution that can be used to test for CO_2 gas.

(v) What is this solution commonly used for in the laboratory? Give a chemical equation for this reaction of the solution.

Explanation:

Limewater is commonly used to test for the presence of CO_2 gas. A white precipitate of insoluble calcium carbonate will form when you bubble CO_2 through limewater:

$$Ca(OH)_2(aq) + CO_2(g) \rightarrow CaCO_3(s) + H_2O(l).$$

 Q Will we get a white precipitate if we bubble CO_2 into aqueous sodium hydroxide solution?

A: When we bubble CO_2 into aqueous hydroxide solution, the following reaction takes place:

$$2OH^-(aq) + CO_2(g) \rightarrow CO_3^{2-}(aq) + H_2O(l).$$

This is an acid-base reaction as the CO_2 is an acidic gas. Thus, insoluble $CaCO_3$ can be obtained because the $CaCO_3$ is insoluble! If CO_2 is bubbled into $NaOH(aq)$, the Na_2CO_3 that is formed is soluble in water. Hence, you would not get a white precipitate! This would mean that if CO_2 is bubbled into aqueous $Ba(OH)_2$, you would get a white precipitate of $BaCO_3$ as well.

Q But when too much CO_2 is bubbled into aqueous $Ca(OH)_2$ solution, no white precipitate is observed. Why?

A: This is because other than forming the carbonate ion, CO_2 can also react with water to form the weak carbonic acid:

$$CO_2(g) + H_2O(l) \rightarrow H_2CO_3(aq).$$

Thus, when too much CO_2 is bubbled into aqueous $Ca(OH)_2$ solution, both the CO_3^{2-} and H_2CO_3 that are formed can react in an acid–base reaction:

$$CO_3^{2-}(aq) + H_2CO_3(aq) \rightarrow 2HCO_3^-(aq).$$

Hence, no white precipitate is formed because $Ca(HCO_3)_2$ is soluble in water.

(vi) Suggest the likely pH of the solution.

Explanation:

The likely pH of the calcium hydroxide solution would be about 12.

 Q What is the best guess of the pH for such a solution?

A: Aqueous ammonia has a pH of about 11. Since $Ca(OH)_2$ is a stronger base than ammonia, the best guess would be a pH value that is more than 11 but less than 13.

(c) Sunglasses can be made from photochromic glass. When bright light strikes photochromic glass, it darkens. This is because photochromic glass contains small amounts of silver chloride, AgCl, and copper(I) chloride, CuCl. In the presence of bright light, silver chloride decomposes into silver atoms which make the glass go dark, and into chlorine atoms:

$$AgCl(s) \rightarrow Ag(s) + Cl(s).$$

Chlorine atoms immediately react with copper(I) chloride to make copper(II) chloride:

$$CuCl(s) + Cl(s) \rightarrow CuCl_2(s).$$

When the exposure to bright light is stopped, silver atoms reduce copper(II) chloride back into copper(I) chloride and silver chloride.

(i) Calculate the maximum mass of silver that can be formed when 0.305 g of silver chloride decomposes.

Explanation:

Molar mass of $AgCl = 107.9 + 35.5 = 143.4$ g mol^{-1}.

Amount of AgCl in 0.305 g $= \dfrac{0.305}{143.4} = 2.13 \times 10^{-3}$ mol.

Amount of Ag formed = Amount of AgCl in 0.305 g $= 2.13 \times 10^{-3}$ mol.

Maximum mass of silver that can be formed $= 2.13 \times 10^{-3} \times 107.9 = 0.229$ g.

Q Why is the state symbol for the Cl atom given as (s)?

A: Well, the state symbol (s) for the Cl atom indicates that it is an atom being trapped within the solid glass matrix.

(ii) Explain why the reaction between copper(I) chloride and chlorine involves both oxidation and reduction.

Explanation:

The oxidation state of Cu^+ in CuCl increases from +1 to +2 in $CuCl_2$. This is an oxidation reaction.

The oxidation state of the Cl atom decreases from 0 to −1 in $CuCl_2$. This is a reduction reaction.

(iii) Give the oxidation and reduction half-equations for the reaction between silver and copper(II) chloride.

Explanation:

Reduction half-equation: $Cu^{2+} + e^- \rightarrow Cu^+$.
Oxidation half-equation: $Ag \rightarrow Ag^+ + e^-$.

(iii) Hence, give the overall ionic equation.

Explanation:

Overall ionic equation: $Cu^{2+} + Ag \rightarrow Cu^+ + Ag^+$.

(iv) Construct the overall equation for the reaction between silver and copper(II) chloride.

Explanation:

When the exposure of light is stopped, we have redox reactions that are reversed of the following two reactions:

$$AgCl(s) \rightarrow Ag(s) + Cl(s);$$

and

$$CuCl(s) + Cl(s) \rightarrow CuCl_2(s).$$

Thus, by reversing the given two equations, we have the following two reactions in sequence:

$$CuCl_2(s) \rightarrow CuCl(s) + Cl(s) \tag{1}$$

$$Ag(s) + Cl(s) \rightarrow AgCl(s). \tag{2}$$

Adding equations (1) and (2), we have:

$$Ag(s) + Cl(s) + CuCl_2(s) \rightarrow AgCl(s) + CuCl(s) + Cl(s)$$

i.e.,

$$Ag(s) + CuCl_2(s) \rightarrow AgCl(s) + CuCl(s).$$

OR

We can simply add two Cl^- ions to each sides of the ionic equation,

$$Cu^{2+} + Ag \rightarrow Cu^+ + Ag^+,$$

to get an electrically neutral equation:

$$Ag(s) + CuCl_2(s) \rightarrow AgCl(s) + CuCl(s).$$

 Q   Wow! Why are you able to add the two chemical equations together like the simultaneous addition of two mathematical equations? What is the reasoning behind such an addition?

A: Well, basically, reaction (1) takes place first. The product in reaction (1), i.e. Cl(s), then participates in reaction (2). So, the overall reaction equation, $Ag(s) + CuCl_2(s) \rightarrow AgCl(s) + CuCl(s)$, seems to indicate that it is the actual reaction between Ag(s) and $CuCl_2(s)$, which is actually not the case. Therefore, you can add chemical equations like that if one product of an equation is also the reactant of another equation!

(v) Aqueous copper(II) chloride reacts with aqueous sodium hydroxide to form a precipitate. Write the ionic equation, including state symbols, for the precipitation reaction.

Explanation:

$$Cu^{2+}(aq) + 2OH^-(aq) \rightarrow Cu(OH)_2(s).$$

(vi) What is the name and color of the precipitate?

Explanation:

The blue precipitate is known as copper(II) hydroxide.

(vii) Give the dot-and-cross diagrams for the precipitate and copper(II) chloride.

Explanation:

$$\left[Cu\right]^{2+} 2\left[:\ddot{\underset{\cdot\cdot}{Cl}}\overset{x}{}\right]^- \quad \left[Cu\right]^{2+} 2\left[:\ddot{\underset{\cdot\cdot}{O}}\overset{x}{}H\right]^-$$

CHAPTER 10

ELECTRIC CELLS AND THE REACTIVITY SERIES

1. (a) A piece of zinc rod was placed in aqueous silver nitrate.

   (i) Write an ionic equation, with state symbols, for the reaction which occurred.

Explanation:

$$Zn(s) + 2Ag^+(aq) \rightarrow Zn^{2+}(aq) + 2Ag(s).$$

(ii) Explain why the reaction involves oxidation and reduction.

Explanation:

The oxidation state of Zn increases from 0 to +2; this is an oxidation process. The oxidation state of Ag^+ decreases from +1 to 0; this is a reduction process.

Q Why is the zinc metal oxidized while the silver ion is reduced?

A: During the collision of two particles, a redox reaction can occur simply because one particle has a higher tendency to *lose* electrons than the other particle. In another perspective, one particle has a higher tendency to *gain* electrons as compared to the other particle.

As different metals have *different potentials* to undergo *oxidation*, how would we know which metal has a higher potential than another? What is the standard of reference?

A: We know that some metals can react with an acid to give hydrogen gas. Hence, we can use hydrogen as a reference to rank the reactivity of different metals. With this, we have the reactivity series:

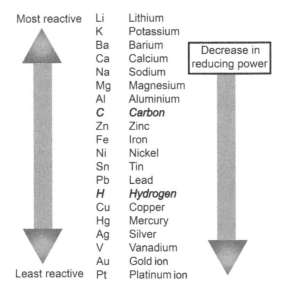

The *reactivity series* ranks the metals according to their ease in undergoing oxidation *relative* to hydrogen. Metals above hydrogen would be able to displace an acid to give hydrogen gas. As we go down the series, the *reducing power* of the metal decreases, i.e., the metals at the lower end of the series are weaker reducing agents as they are less likely to undergo oxidation.

Q   Why is hydrogen used as the reference instead of H^+?

A: Now, our purpose is to compare which metal has a higher potential to undergo oxidation. When a metal is oxidized, it is acting as a reducing agent. Therefore, the greater the potential of the metal to undergo oxidation, the stronger the reducing power!

It is inappropriate to compare a metal with H^+ because H^+ cannot be further oxidized. H^+ can only be reduced and when it is being reduced, it is acting as an oxidizing agent. So, how can you compare a reducing agent (which is the metal) with an oxidizing agent (which is H^+)? This would be like comparing an apple with an orange — invalid!

Hence, if a metal can react with an acid to give hydrogen gas, then it must be a *stronger* reducing agent than hydrogen.

$$Mg(s) \quad + \quad 2H^+(aq) \quad \rightarrow \quad Mg^{2+}(aq) \quad + \quad H_2(g).$$
Reducing Agent 1   Oxidizing Agent 2   Oxidizing Agent 1   Reducing Agent 2

Thus, if the above reaction occurs, we can conclude that:

- Reducing Agent 1 (Mg) is stronger than Reducing Agent 2 (H_2). Therefore, Mg is above hydrogen in the reactivity series.
- Oxidizing Agent 2 (H^+) is stronger than Oxidizing Agent 1 (Mg^{2+}).

Did you notice that the forward and backward reactions are in fact a redox reaction?

(iii) What would you expect to see after the reaction has taken place for some time?

Explanation:

The piece of zinc metal would slowly dissolve with silvery deposits, which is the silver metal, coated on its surface.

(b) Sodium is stored under oil because it rapidly oxidizes to form sodium oxide, Na_2O, a base. Sodium oxide reacts with water to form sodium hydroxide, which is an alkali.

(i) Explain the terms *base* and *alkali*.

Explanation:

A base is a substance that reacts with an acid to give a salt and water.

A soluble base is called an alkali and it dissolves in water to produce the hydroxide ion, OH^-.

Do you know?

— A base is the chemical opposite of an acid. A reaction between an acid and a base is termed *neutralization*, producing a salt and water. Other than that, bases can also react with ammonium salts to form a salt, water, and ammonia:

$$KOH(aq) + NH_4Cl(aq) \rightarrow KCl(aq) + H_2O(l) + NH_3(g).$$

This can serve as a characteristic test for the presence of NH_4^+ ions. In addition, alkalis react with some salts to produce insoluble hydroxides:

$$2NaOH(aq) + Cu(NO_3)_2(aq) \rightarrow Cu(OH)_2(s) + 2NaNO_3(aq).$$

The color of the precipitate can serve as characteristic identification of the cation that is present in the unknown solution.

(ii) Write the equations for the above two reactions.

Explanation:

$$4Na(s) + O_2(g) \rightarrow 2Na_2O(s);$$

and

$$Na_2O(s) + H_2O(l) \rightarrow 2NaOH(aq).$$

(iii) Give the dot-and-cross diagram for sodium oxide.

Explanation:

$$2\left[Na\right]^{+}\left[:\overset{\bullet\bullet}{\underset{\bullet \times}{O}} {}_{\times}\right]^{2-}$$

Q Can we draw the following dot-and-cross diagram instead?

$$2\left[Na\right]^{+}\left[:\overset{\bullet\bullet}{\underset{\times\times}{O}}:\right]^{2-}$$

A: Yes! As long as you indicate that the two electrons which the oxygen atom has gained with a different symbol from those electrons that are originally present in the valence shell. But if you are interested to find out which is preferred, you can refer to *Understanding Advanced Physical Inorganic Chemistry* by J. Tan and K.S. Chan.

(iv) The reaction between sodium and oxygen is in fact a redox reaction. Give both the oxidation and reduction half-equations.

Explanation:

Oxidation half-equation: $Na \rightarrow Na^+ + e^-$.

Reduction half-equation: $O_2 + 4e^- \rightarrow 2O^{2-}$.

Do you know?

— Oxygen is a common component in air, thus the reactivity of a metal with oxygen would decide whether the metal can exist in the pure metallic form or the oxide form in nature. Based on the reactivity series, we would expect that the more reactive metals will be less likely to exist in their pure metallic forms in the presence of air.

— Highly reactive metals such as K and Na form the metal oxides in a limited amount of oxygen:

$$4K(s) + O_2(g) \rightarrow 2K_2O(s);$$

and

$$4Na(s) + O_2(g) \rightarrow 2Na_2O(s).$$

But in excess oxygen, these two metals form the peroxides (O_2^{2-}), which are important oxidizing agents in the bleaching of wood pulp for the production of paper and textiles:

$$2K(s) + O_2(g) \rightarrow K_2O_2(s);$$

and

$$2Na(s) + O_2(g) \rightarrow Na_2O_2(s).$$

— The rest of the metals in the reactivity series, such as Ca, Mg, Al, Zn, and Fe, burn with decreasing reactivity to form the oxides:

$$2M(s) + O_2(g) \rightarrow 2MO(s), \quad M = Ca, Mg, Zn, and Fe.$$

$$e.g., 4Al(s) + 3O_2(g) \rightarrow 2Al_2O_3(s).$$

Silver and gold do not oxidize in air and this is probably the main reason why these two metals have already been widely used in their metallic forms in ancient times.

— The reactivity of the metal with oxygen decreases in the order:

$$K > Na > Ca > Mg > Al > Zn > Fe > Pb > Cu > Ag > Au,$$

which is GENERALLY in accordance to their ranking in the reactivity series!

> **Q** Why does K and Na form peroxides with oxygen whereas the rest of the metals form the normal oxide?

A: This has got to do with the charge density $\left(\alpha \dfrac{q_+}{r_-} \right)$ of the metal cation. Both Na^+ and K^+ are singly charged, whereas Ca^{2+}, etc., are doubly positively charged. The high charge density of these doubly charged cations would polarize the electron cloud of the peroxide anion (O_2^{2-}) to the extent that it becomes highly unstable. Hence, these doubly charged cations "prefer" to form the oxide instead.

> **Q** Wait a minute, lithium ion (Li^+) is also singly charged. Why does it not form Li_2O_2 (lithium peroxide)?

A: Well, this still falls back on the concept of charge density. Lithium ion, though singly charged, has a very small ionic radius. Therefore, its charge density is high, leading to a high polarizing power.

(v) 62 g of sodium oxide is used to make 1.0 dm^3 of aqueous sodium hydroxide. What is the concentration, in mol dm^{-3}, of the sodium hydroxide solution?

Explanation:

$$Na_2O(s) + H_2O(l) \rightarrow 2NaOH(aq).$$

Amount of Na_2O in 62 g $= \dfrac{62}{62.0} = 1$ mol.

Amount of NaOH $= 2 \times$ Amount of Na_2O in 62 g $= 2$ mol.

Concentration of the sodium hydroxide solution $= \dfrac{2}{1} = 2$ mol dm^{-3}.

> **Q**   Is the reaction between $Na_2O(s)$ and $H_2O(l)$ an acid–base reaction?

A: Yes, indeed! The base is the O^{2-} ion while the acid is the H_2O molecule.

> **Q**   If it is an acid–base reaction, why is there no salt and water formed?

A: Well, this is an example of an acid–base reaction in which a salt (NaOH) is formed but no water is formed. Thus, an acid–base reaction giving a salt and water is just a guideline! More importantly, you need to understand in an acid–base reaction, the main concept is the transfer of a H^+ ion from an acid to a base!

> (vi) Determine the volume of $0.50 \, mol \, dm^{-3}$ sulfuric(VI) acid that is required to neutralize $20 \, cm^3$ of the sodium hydroxide solution.

Explanation:

$$H_2SO_4(aq) + 2NaOH(aq) \rightarrow Na_2SO_4(aq) + 2H_2O(l).$$

Amount of NaOH in $20 \, cm^3 = \dfrac{20}{1000} \times 2.0 = 0.04$ mol.

Amount of H_2SO_4 needed $= \dfrac{1}{2} \times$ Amount of NaOH in $20 \, cm^3 = \dfrac{1}{2} \times 0.04 = 0.02$ mol.

Volume of $0.50 \, mol \, dm^{-3}$ H_2SO_4 needed $= \dfrac{0.02}{0.50} = 0.04 \, dm^3 = 40 \, cm^3$.

> 2. (a) Chromium lies between zinc and iron in the reactivity series. In its simple chemistry, chromium has a valency of 3 and forms green compounds.
>
>   (i) State which particle in the nucleus is responsible for the element's position in the Periodic Table.

Explanation:

The particle in the nucleus that is responsible for the element's position in the Periodic Table is the proton.

Do you know?

— The reactivity series informs us about the relative *reducing power* of the metal, which is important to help us understand the reactivities of different metals with substances such as an acid, carbon, hydrogen, and even other metals. It is also important to take note that if the *metal* itself is acting as a *reducing agent* by losing electrons, then the *metal ion* must be an *oxidizing agent*, which gains electrons. This would mean that if a metal "loves" to lose electrons, then its corresponding cation would "hate" to take in electrons.

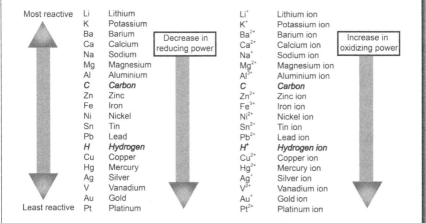

Thus, the *oxidizing power* of the *metal cation increases* down the series, contrary to the decreases of the reducing power of the metal.

(ii) State which particle in the nucleus is responsible for the presence of isotopes.

Explanation:

The particle in the nucleus that is responsible for the presence of isotopes is the neutron.

> (iii) Explain what you will observe if an excess amount of powdered chromium is added to aqueous copper(II) sulfate.

Explanation:

Chromium is above copper in the reactivity series. This means that chromium metal is more likely to undergo oxidation (lose electrons) than the copper metal. Hence, if an excess amount of powdered chromium is added to aqueous copper(II) sulfate, the following redox reaction would take place:

$$2Cr(s) + 3Cu^{2+}(aq) \rightarrow 2Cr^{3+}(aq) + 3Cu(s).$$

Thus, we would observe the dissolution of the chromium powder with brown deposits of copper metal formed.

Do you know?

— As we go down the reactivity series, the reducing power of the metal decreases. This means that the more reactive metals can actually reduce the less reactive metals from their oxides or salt solutions. Such is the basis behind the *displacement* reaction. We have seen one above:

$$2Cr(s) + 3Cu^{2+}(aq) \rightarrow 2Cr^{3+}(aq) + 3Cu(s).$$

Other examples include:

$$Fe_2O_3(s) + 2Al(s) \rightarrow Al_2O_3(s) + 2Fe(s); \quad \text{(the Thermite reaction)}$$

$$Fe(s) + Cu^{2+}(aq) \rightarrow Fe^{2+}(aq) + Cu(s); \quad \text{and}$$

$$Cu(s) + 2Ag^+(aq) \rightarrow Cu^{2+}(aq) + 2Ag(s).$$

In a nutshell, the more reactive metal forms compounds more readily than the less reactive metal.

What is the Thermite reaction used for?

A: The Thermite reaction helps in welding railway tracks together. It is highly exothermic in nature, a demonstration that $Al_2O_3(s)$ and $Fe(s)$ are energetically more stable than $Fe_2O_3(s)$ and $Al(s)$.

(iv) Chromium can be manufactured by the chemical reduction of chromium(III) oxide. Suggest a suitable reagent and the probable condition for this reduction.

Explanation:

Chromium is below carbon in the reactivity series. This means that carbon is more likely to undergo oxidation (lose electrons) than chromium metal. Hence, if chromium(III) oxide is mixed with carbon, the following redox reaction would take place:

$$2Cr_2O_3(s) + 3C(s) \rightarrow 4Cr(s) + 3CO_2(g).$$

Thus, we can use carbon to reduce chromium(III) oxide at a temperature of about 1000°C.

 How do you know to state the temperature as 1000°C?

A: Well, the reduction of iron(III) oxide in the blast furnace by carbon needs a temperature of about 900–1100°C. Now, since we were told that "chromium lies between zinc and iron in the reactivity series," to say the temperature needed is about 1000°C would be a good guess. This is a skill for you to master — to answer questions based on what you have learned!

Do you know?

— The reactivity series of metal informs us how easily it is for the metal to undergo oxidation, i.e., to be a reducing agent. We have learned that the easier the metal is able to be oxidized, the more difficult it is for the corresponding cation to be reduced. Hence, we would expect the metal oxides for highly reactive metals to be more difficult to decompose, i.e., to be reduced back to its corresponding metal.

— Metal oxides of K, Na, Ca, Mg, and Al, which are *above carbon* in the reactivity series would not be able to be reduced by carbon back to its metallic form. Oxides of Zn, Fe, Pb, and Cu would be able to be reduced by carbon through heating:

$$2MO(s) + C(s) \rightarrow 2M(s) + CO_2(g). \quad M = Zn, Fe, Pb, \text{ and } Cu.$$

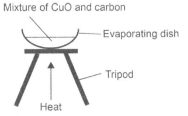

Mixture of CuO and carbon

Evaporating dish

Tripod

Heat

— If the metal is below Cu in the reactivity series, its corresponding metal oxide can simply be reduced via heating:

$$2Ag_2O(s) \rightarrow 4Ag(s) + O_2(g).$$

— Knowing all these is important because it would help us know how to extract the metal from its ore. Oxides above carbon in the reactivity series have to be extracted via electrolysis, while oxides below carbon can be extracted simply by reducing it with carbon.

(v) Chromium is usually used to plate iron to prevent it from rusting. Explain how iron coated with chromium is protected from rusting.

Explanation:

Chromium is above iron in the reactivity series; this means that chromium would be more likely undergo oxidation compared to iron. The layer of chromium(III) oxide that is formed can prevent the iron underneath it from rusting as it is impervious to water and oxygen.

Do you know?

— The fact that you have oxygen in iron oxide means that iron and oxygen will react! The reaction between iron and oxygen is a redox reaction in which the iron undergoes oxidation while the oxygen is reduced. When cast iron (iron containing impurities) comes in contact with water and oxygen, it corrodes or in layman's terms, we say it rusts. Iron in the absence of either water or oxygen would never corrode. If the water contains other electrolytes such as a salt or an acid, rusting is accelerated. That is why iron corrodes faster in seawater.

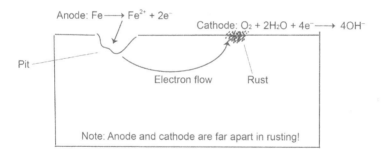

The brown rust that is formed is actually a form of hydrated iron(III) oxide:

$$4Fe(s) + 3O_2(g) + 2xH_2O(l) \rightarrow 2Fe_2O_3 \cdot xH_2O(s).$$

The number of water of crystallization depends very much on the reaction conditions in which the rust forms.

(vi) Explain why acid is detrimental to chromium as a protector of iron. Give a chemical equation for the reaction between chromium(III) oxide and hydrochloric acid.

Explanation:

Chromium(III) oxide is basic in nature. Thus, it can react with acid through an acid–base reaction. With the dissolution of the chromium(III) oxide, the protector of iron is gone and the iron can undergo rusting:

$$Cr_2O_3(s) + 6HCl(aq) \rightarrow 2CrCl_3(aq) + 3H_2O(l).$$

 Q Water is also an acid. But why did it not react with the basic chromium(III) oxide?

A: Good question! Water is an acid but unfortunately, it is not strong enough to react with the basic chromium(III) oxide. So, the key takeaway here is: an acid cannot react with all bases, and neither can a base react with all kinds of acids!

 Q But how can we explain that water can react with Na_2O but not with oxides such as Cr_2O_3 or Al_2O_3?

A: Well, this is not difficult. In order to react, the oxide must dissolve first! For oxides such as Cr_2O_3 or Al_2O_3 and many others, if they cannot dissolve to release the O^{2-} ion, then the O^{2-} ion cannot react with the weakly acidic H_2O molecule. So, Na_2O is able to react with water molecules probably because of the weak ionic bonds in Na_2O, which can be easily overcome.

So, Cr_2O_3 or Al_2O_3 and many others cannot react with water because they are not very soluble in water. How would you account for the reaction of the oxides with hydrochloric acid?

A: In aqueous hydrochloric acid, there are free H^+ ions. These ions can attack the O^{2-} ions on the surfaces of the ionic solid more readily than the H_2O molecules because the H^+ ion is positively charged. The H atoms of the H–O–H molecule are only slightly electron- deficient, i.e., possessing a partial charge (δ^+). Thus, they are not readily attracted to the O^{2-} ions.

Wow! I did not know that a chemical reaction such as the reaction of an oxide with acid can be so complicated!

A: Actually, it is not really complicated! In a chemical reaction, it can be divided into three stages: (1) the attacking stage; (2) bond-breaking stage; and (3) bond-forming stage. Sometimes, different factors dominating each of the three stages can decide whether the reaction would proceed or not. For instance, the reaction of Na_2O with H_2O is feasible because of a greater ease in the bond-breaking process. For the reaction of Cr_2O_3 and HCl, the greater ease in the attacking stage causes the reaction to proceed. Thus, the learning of chemistry is very frustrating to many students because they fail to appreciate the need to apply the concepts that they have learned contextually!

Is there an enjoyable way to learn chemistry?

A: You need to understand the basic concepts well! First, know the limitations and whys of the concept. Secondly, always use your imagination to visualize how a reaction is happening: what is going on during the reaction, which particle is attacking which, what bonds are breaking, and what bonds are forming. Thirdly, list down the factors that may be affecting the reaction. Finally, identify the factors that can explain the observation or data that you have obtained. All these tips would help you to enjoy your learning of chemistry.

(b) Concentrated hydrochloric acid can be used to remove rust marks from porcelain sinks and baths.

(i) Write a balanced chemical equation for the reaction between rust, assumed to be iron(III) oxide, (Fe_2O_3), and hydrochloric acid.

Explanation:

$$Fe_2O_3(s) + 6HCl(aq) \rightarrow 2FeCl_3(aq) + 3H_2O(l).$$

(ii) The concentration of an industrially produced hydrochloric acid solution is determined by titrating against a standard solution of sodium hydroxide. The concentrated hydrochloric acid is diluted 50 times. Then, $10\,cm^3$ of this diluted solution was titrated against $9.2\,cm^3$ of $0.15\,mol\,dm^{-3}$ sodium hydroxide solution. Calculate the concentration in g dm^{-3} of the concentrated hydrochloric acid.

Explanation:

$$NaOH(aq) + HCl(aq) \rightarrow NaCl(aq) + H_2O(l).$$

Amount of NaOH used for titration $= \dfrac{9.2}{1000} \times 0.15 = 1.38 \times 10^{-3}\,mol.$

Amount of HCl in $10\,cm^3$ of diluted solution = Amount of NaOH used for titration $= 1.38 \times 10^{-3}\,mol.$

Concentration of diluted HCl $= 1.38 \times 10^{-3} \div \dfrac{10}{1000} = 0.138\,mol\,dm^{-3}.$

Since the diluted HCl is obtained by diluting 50 times the original concentrated HCl, the original concentration of the concentrated HCl $= 50 \times 0.138 = 6.9\,mol\,dm^{-3}.$

(c) 1.267 g of a metal, **X**, was placed in a test tube and an excess of hot silver(I) nitrate solution was added to it and stirred until the reaction stopped. The metal **X** displaced silver from the silver(I) nitrate solution. The silver was filtered off, washed with water, and dried in an oven at 100°C. The dry silver had a mass of 4.30 g. The reaction that occurred is represented by the equation:

$$\mathbf{X}(s) + 2AgNO_3(aq) \rightarrow \mathbf{X}(NO_3)_2(aq) + 2Ag(s).$$

(i) Explain why a hot solution of silver(I) nitrate was used.

Explanation:

The reaction between **X** and Ag^+ is probably slow at room temperature. A higher temperature would speed up the reaction.

(ii) Explain why an excess of silver(I) nitrate was used.

Explanation:

As the amount of **X**, i.e., the number of moles, is not known, an excess of silver(I) nitrate would ensure that all the **X** have reacted.

(iii) Suggest why the silver was dried at 100°C rather than at a higher temperature.

Explanation:

Silver can already react with oxygen at room temperature. At a higher temperature, the silver will react faster with oxygen in the air to give silver(I) oxide (Ag_2O). In addition, the boiling point of water is 100°C; thus, we do not need a temperature that is higher than 100°C to vaporize off the water.

(iv) Calculate the number of moles of silver that were formed and hence the number of moles of **X** that were used in the reaction.

Explanation:

Amount of Ag in $4.30\,g = \dfrac{4.30}{107.9} = 3.99 \times 10^{-2}$ mol.

Amount of **X** used $= \dfrac{1}{2} \times$ Amount of Ag in $4.30\,g = 1.99 \times 10^{-2}$ mol.

(v) Calculate the relative atomic mass of **X**. Use the Periodic Table to identify the metal **X**.

Explanation:

Let the molar mass of **X** be $x\ \mathrm{g\,mol^{-1}}$.

Amount of **X** used $= 1.99 \times 10^{-2}$ mol $= \dfrac{1.267}{x}$

$$\Rightarrow x = \dfrac{1.267}{0.0199} = 63.7 \text{ g mol}^{-1}.$$

Therefore, the relative atomic mass of **X** is 63.7.

From the Periodic Table, **X** is probably Cu.

Q   Why is Cu able to reduce Ag^+?

A: Copper is above silver in the reactivity series. This would mean that copper is more likely to undergo oxidation (lose electrons) than silver metal. Hence, if copper is mixed with Ag^+, the following redox reaction (displacement) would take place:

$$Cu(s) + 2Ag^+(aq) \rightarrow Cu^{2+}(aq) + 2Ag(s).$$

(vi) Explain, in terms of electron transfer, why the reaction can be considered to involve both oxidation and reduction.

Explanation:

The oxidation state of Cu metal increases from 0 to +2; this is an oxidation reaction. The oxidation state of Ag^+ decreases its oxidation state from +1 to 0; this is a reduction reaction.

(vii) Give both the anode and cathode reactions.

Explanation:

Anode (oxidation) reaction: $Cu \rightarrow Cu^{2+} + 2e^-$.

Cathode (reduction) reaction: $Ag^+ + e^- \rightarrow Ag$.

Do you know?

— In electrochemistry, an *anode* simply means the *oxidation* electrode while *cathode* implies the *reduction* electrode. The anode or cathode is NOT determined by its polarity or sign.

The *anode*:

The more reactive metal undergoes *oxidation*, therefore the metal electrode is known as the *anode*. It is *negatively* charged because electrons are deposited onto it during oxidation.

$$M \rightarrow M^{n+} + ne^-.$$

The *cathode*:

At the less reactive metal, ions from the electrolyte take up the electrons flowing from the anode and undergoes *reduction*. This metal electrode

(*Continued*)

(Continued)

is thus known as the *cathode*. The electrode is *positively* charged because after the cation has "consumed" the electrons, it "deposits its positive charge" onto the electrode.

$$L^{y+} + ye^- \rightarrow L.$$

(viii) Give the overall ionic equation for the reaction.

Explanation:

Overall ionic equation for the reaction:

$$Cu(s) + 2Ag^+(aq) \rightarrow Cu^{2+}(aq) + 2Ag(s).$$

3. Magnesium metal is above hydrogen in the reactivity series. Dilute ethanoic acid and dilute hydrochloric acid both react with magnesium ribbon to form hydrogen.

(a) Give the electronic configuration of the magnesium atom.

Explanation:

The electronic configuration of the magnesium atom is 2.8.2.

(b) Give the formula of one ion found in both of these dilute acids.

Explanation:

The formula of one ion found in both of these dilute acids is $H^+(aq)$ or $H_3O^+(aq)$.

Do you know?

— Hydrogen ion does not exist in solution. H^+ ion has such a high charge density that it is actually bonded to at least one water molecule when in aqueous solution i.e., the H^+ ion binds with H_2O to form H_3O^+ (hydronium ion).

Thus, without the water molecule, HCl or ethanoic acid cannot function as an acid!

(c) Magnesium ribbon reacts with hydrochloric acid as shown in the equation:

$$Mg(s) + 2HCl(aq) \rightarrow MgCl_2(aq) + H_2(g).$$

A 0.24 g sample of magnesium ribbon is added to 5.0 cm³ of 2.0 mol dm⁻³ hydrochloric acid.

(i) Draw the dot-and-cross diagram for magnesium chloride.

Explanation:

(ii) Which reactant, magnesium or hydrochloric acid, is in excess? Use calculations to explain your answer.

Explanation:

Amount of Mg in 0.24 g of magnesium ribbon $= \dfrac{0.24}{24.3} = 9.88 \times 10^{-3}$ mol.

Amount of HCl in 5.0 cm^3 $= \dfrac{50}{1000} \times 2.0 = 1.00 \times 10^{-2}$ mol.

From the equation, 9.88×10^{-3} moles of Mg would need $2 \times 9.88 \times 10^{-3} = 1.98 \times 10^{-2}$ moles of HCl.

Hence, magnesium is in excess.

> (iii) Calculate the maximum mass of magnesium chloride that can be formed in this reaction.

Explanation:

Amount of MgCl$_2$ formed $= \dfrac{1}{2} \times$ Amount of HCl in 5.0 cm^3 $= \dfrac{1}{2} \times 1.00 \times 10^{-2} = 5.00 \times 10^{-3}$ mol.

Molar mass of MgCl$_2$ = 24.3 + 2(35.5) = 95.3 g mol^{-1}.

Maximum mass of magnesium chloride that can be formed $= 5.00 \times 10^{-3} \times 95.3 = 0.477$ g.

> (iv) A 0.24 g sample of magnesium ribbon is added to 5.0 cm^3 of 2.0 mol dm^{-3} ethanoic acid. Explain why this reaction forms the same volume of hydrogen, but takes place much more slowly than the reaction of the same mass of magnesium with 5.0 cm^3 of 2.0 mol dm^{-3} hydrochloric acid.

Explanation:

Amount of ethanoic acid in 5.0 cm^3 $= \dfrac{5.0}{1000} \times 2.0 = 1.00 \times 10^{-2}$ mol = Amount of HCl in 5.0 cm^3.

Since the amount of ethanoic acid is the same as that of HCl and we are using the same amount of Mg, the same volume of H_2 will be obtained.

Ethanoic acid is a weak acid that does not fully dissociate in water. Since the concentration of free H^+ ions in ethanoic acid is lower than that in HCl, the rate of reaction of the ethanoic acid and magnesium is slower.

(v) Write an equation for the reaction between dilute ethanoic acid and sodium carbonate. What observations would be made during this reaction?

Explanation:

$2CH_3COOH(aq) + Na_2CO_3(s) \rightarrow 2CH_3COONa(aq) + CO_2(g) + H_2O(l).$

Effervescence of CO_2 gas would be observed.

(vi) Suggest a test for your observation in part (v).

Explanation:

Bubble the gas that is evolved into aqueous calcium hydroxide solution. A white precipitate of $CaCO_3$ would be observed.

(vii) Draw the dot-and-cross diagram for the gas in part (v).

Explanation:

$$\overset{\times\times}{\underset{\times\times}{O}} {\times}{\colon} C {\colon}{\times} \overset{\times\times}{\underset{\times\times}{O}} {\times}$$

(d) When a piece of magnesium metal and a piece of copper metal were dipped in a solution of copper(II) sulfate, electrons flowed from the magnesium electrode to the copper electrode. A brown deposit was found on the surface of the copper metal. The magnesium metal was acting as the anode, while the copper electrode was the cathode.

(i) Explain the terms *anode* and *cathode*.

Explanation:

An *anode* means the *oxidation* electrode while *cathode* implies the *reduction* electrode.

(ii) Give the half-equations at both the anode and cathode.

Explanation:

Anode (oxidation) reaction: $Mg \rightarrow Mg^{2+} + 2e^-$.

Cathode (reduction) reaction: $Cu^{2+} + 2e^- \rightarrow Cu$.

(iii) Give the overall reaction equation that has taken place.

Explanation:

Overall reaction equation: $Mg(s) + Cu^{2+}(aq) \rightarrow Cu(s) + Mg^{2+}(aq)$.

Q   Why is Mg able to reduce Cu^{2+}?

A: Magnesium is above copper in the reactivity series. This means that magnesium is more likely to undergo oxidation (lose electrons) than copper metal. Hence, if magnesium comes in contact with Cu^{2+}, the above redox reaction (displacement) would take place.

(iv) Give the polarities of both electrodes.

Explanation:

The anode is negatively charged because electrons are deposited onto it during oxidation. The cathode is positively charged because after the cation has "consumed" the electrons, it "deposits its positive charge" onto the electrode.

(v) Suggest the role played by the copper(II) sulfate solution. Give a name of copper(II) sulfate in this set-up.

Explanation:

The copper(II) sulfate is the electrolyte in the system. An electrolyte is a substance containing free ions that make the substance electrically conductive.

Do you know?

— An electric cell is a type of *electrochemical* cell that converts *chemical* energy into *electrical* energy. It is constructed by immersing two metal

(Continued)

(Continued)

electrodes, of different reactivities, into an electrolyte containing mobile ions that can function as charge carriers.

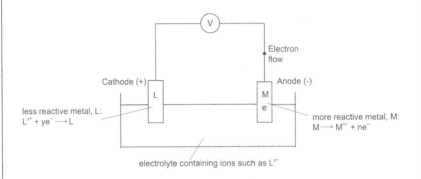

When a wire connects the two electrodes, electrons flow from the *more reactive* metal (the anode) through the wire of the external circuit to the *less reactive* metal (the cathode).

 Q What causes the electrons to flow from the more reactive metal to the less reactive one?

A: It is due to the difference in the *ability* to undergo oxidation or lose electrons. This difference is technically called a *potential difference* or *electromotive force* (e.m.f.).

 Q What would happen to the more reactive metal after oxidation?

A: It forms the cation and *goes into the solution* to maintain electrical neutrality as over at the cathode, the cation undergoes discharge to form the neutral atom. Thus, with the discharge of the cation, the electrolyte would become increasingly negatively charged because of the remaining anions, which can be diminished by the cation that is formed during oxidation.

Q   What would happen to the more reactive metal after some time?

A: It would shrink in size. As time passes, if one of the species is depleted, the electric cell cannot function any longer. We won't be able to tap any more electrical energy from it. It is like a battery going flat!

(vi) What would happen if the copper metal is replaced witwh a piece of calcium metal?

Explanation:

Assuming that the calcium metal does not react with the water in the electrolyte, electrons will flow from the calcium electrode to the magnesium electrode. A brown deposit would be found on the surface of the magnesium metal.

Q   Why is the copper deposited onto the magnesium instead?

A: Calcium is above magnesium and copper in the reactivity series. This means that calcium is more likely to undergo oxidation (lose electrons) than magnesium metal. Hence, the electrons would flow from the calcium electrode to the magnesium electrode. Since only the Cu^{2+} ions can undergo reduction in the system, the Cu^{2+} ions would gain the electrons and be deposited as copper metal onto the magnesium electrode.

Q   You said, "Assuming that the calcium metal does not react with the water in the electrolyte," does that mean that calcium metal can react with water?

A: Yes, water can function as an acid and react with calcium metal. But the rates of reaction of the metals with water really depend on the reactivities of the metals. In fact, the reactivity of the metal with water or acid decreases in the order:

$$K > Na > Ca > Mg > Al > Zn > Fe > Pb > Cu > Ag > Au.$$

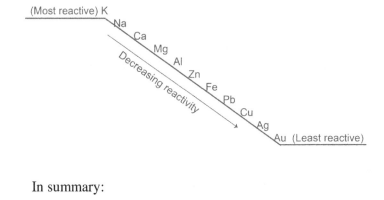

In summary:

Metal	Reaction with cold water	Reaction with steam	Reaction with dilute strong acid
K	Reacts with cold water in decreasing reactivity (K, very violently; Mg, very slowly). The corresponding hydroxide and hydrogen gas are formed. For example: $2K(s) + 2H_2O(l) \rightarrow 2KOH(aq) + H_2(g)$ $Mg(s) + 2H_2O(l) \rightarrow Mg(OH)_2(s) + H_2(g)$	Reacts with steam in decreasing reactivity (K, very violently; Fe, very slowly and need to be red hot). The corresponding metal oxide is formed instead of the metal hydroxide. For example: $2Al(s) + 3H_2O(g) \rightarrow Al_2O_3(s) + 3H_2(g)$	Reacts with dilute strong acids in decreasing reactivity (K, very explosively; Al, rapidly; Zn, moderately fast; Fe, slowly; Pb, very slowly). For example: $2Na(s) + 2HCl(aq) \rightarrow 2NaCl(aq) + H_2(g)$ $Zn(s) + 2HCl(aq) \rightarrow ZnCl_2(aq) + H_2(g)$
Na			
Ca			
Mg			
Al	Does not react with cold water!		
Zn			
Fe			
Pb		Does not react with steam!	
Cu			Does not react with dilute strong acids!
Ag			
Au			

Why is the metal oxide formed instead of the metal hydroxide when the metal reacts with steam?

A: When the metal reacts with steam, the *temperature* is high. Hence, the metal hydroxide undergoes dehydration to form the metal oxide:

$$M(OH)_2(s) \rightarrow MO(s) + H_2O(l).$$

 If Al can react with steam, how can we use Al metal for cookware?

A: The highly insoluble Al_2O_3 that is formed would prevent the Al metal beneath from further reaction. This is known as chemical passivation, as the metal becomes "passive" or unreactive.

(vii) If the mass of the brown deposit is 2.0 g in one hour, calculate the change in mass of the magnesium electrode.

Explanation:

Amount of Cu (brown deposit) = $\dfrac{2.0}{63.5}$ = 3.15×10^{-2} mol.

Amount of Mg oxidized = Amount of Cu (brown deposit) = 3.15×10^{-2} mol.

Mass of Mg oxidized = $3.15 \times 10^{-2} \times 24.3 = 0.765$ g.

(viii) Give the electronic configuration of the copper(II) ion.

Explanation:

The electronic configuration of the copper(II) ion is 2.8.17.

(ix) In view of the position of magnesium in the reactivity series, suggest how we can get magnesium metal from its ore.

Explanation:

Magnesium is above carbon in the reactivity series; thus, magnesium can only be obtained through the electrolysis of its ore.

 So, will the method of obtaining magnesium from its ore make the metal expensive?

A: Yes, certainly. Electricity is more expensive than the burning of carbon. Metals such as magnesium and aluminum, which are obtained through electrolytic means, are very expensive.

CHAPTER 11

ELECTROLYSIS

Do you know?

— In an electric cell, chemical energy is converted into electrical energy by directing the electron transfer during a redox reaction, through an external circuit. But in the electrolytic cell, an electric current is passed through an electrolyte to force an otherwise *non-spontaneous* redox reaction to occur. This process is known as *electrolysis* and the set-up is termed the *electrolytic cell*.

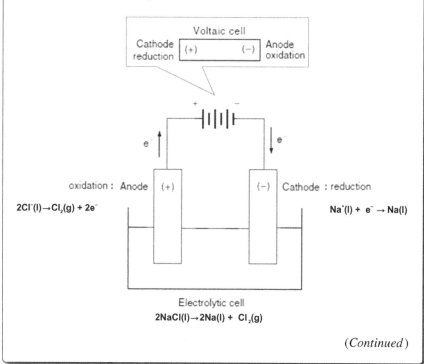

(Continued)

(Continued)

— In the electrolytic cell, we have:

- The *electrolyte*, which is a compound in *solution* or a *molten* compound that is able to conduct electricity because there are *mobile* charge carriers in the form of ions. The electrolyte is decomposed in the process.
- An electric cell or power source, which is used to drive the flow of electrons in the electrolytic set-up. It acts like a "pump" to push electrons in a single direction.
- The *electrodes*, which are metal conductors or graphite, by which an electric current enters or leaves the electrolyte.
- The *cathode*, which is the *reduction* electrode and it is *negatively* charged because it is *connected* to the *negative* terminal of the electric cell. Take note that the polarity or sign of the cathode for an electrolytic cell is *opposite* of that for an electric cell.
- The *anode*, which is the *oxidation* electrode and it is *positively* charged because it is *connected* to the *positive* terminal of the electric cell. Take note that the polarity or sign of the anode for an electrolytic cell is *opposite* of that for an electric cell.

1. Electrolysis involves the decomposition of a compound by the passage of an electric current.

(a) Complete the table, which relates to the electrolysis of different solutions using inert electrodes.

Electrolyte	Ions present in electrolyte	Product at anode	Product at cathode
Molten potassium chloride			
Dilute aqueous potassium nitrate	K^+, H^+, OH^-, and NO_3^-	oxygen	
Concentrated aqueous sodium chloride	Na^+, H^+, OH^-, and Cl^-	chlorine	hydrogen
Dilute aqueous copper(II) sulfate	Cu^{2+}, SO_4^{2-}, H^+, and OH^-		copper
Dilute sulfuric acid		oxygen	hydrogen

Explanation:

Electrolyte	Ions present in electrolyte	Product at anode	Product at cathode
Molten potassium chloride	K^+, and Cl^-	chlorine	potassium
Dilute aqueous potassium nitrate	K^+, H^+, OH^-, and NO_3^-	oxygen	hydrogen
Concentrated aqueous sodium chloride	Na^+, H^+, OH^-, and Cl^-	chlorine	hydrogen
Dilute aqueous copper(II) sulfate	Cu^{2+}, SO_4^{2-}, H^+, and OH^-	*oxygen*	copper
Dilute sulfuric acid	H^+, OH^-, and SO_4^{2-}	oxygen	hydrogen

Q Why are there OH^- ions present in dilute sulfuric acid?

A: Water can undergo auto-ionization:

$$H_2O(l) + H_2O(l) \rightleftharpoons H_3O^+(aq) + OH^-(aq).$$

Thus, both $H^+(aq)$ and $OH^-(aq)$ ions are present in the aqueous solution irrespective of whether it is an acidic or alkaline solution. In an acidic solution, $[H^+(aq)]$ is greater than $[OH^-(aq)]$; vice versa in an alkaline solution.

Do you know?

— In Chapter 10, we learned about the reactivity series, which tells us that different metals have different potentials to undergo oxidation. At the same time, we also learned that the greater the ease of a metal in undergoing oxidation, its correspondng metal cation would be less likely to

(Continued)

(Continued)

be reduced. With this, we have the following diagram to show us the relative ease of discharge of the various cations and anions:

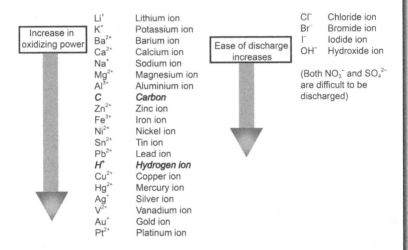

This diagram would help us decide which anion would undergo discharge during electrolysis. Students need to remember that both SO_4^{2-} and NO_3^- would have difficulty undergoing discharge in an aqueous solution. Instead, the *predominant* discharge to replace the discharge of both the SO_4^{2-} and NO_3^- would be the hydroxide ion that is present:

$$4OH^-(aq) \rightarrow 2H_2O(l) + O_2(g) + 4e^-.$$

(i) Explain the terms *inert electrode*, *anode*, and *cathode*.

Explanation:

An inert electrode is an electrode that does not react with the components inside the electrolytic cell. An anode is the electrode in which oxidation takes place while the cathode is where reduction occurs.

Q Why is graphite preferred to platinum as an inert electrode?

A: Well, because it is cheap!

(ii) Explain the source of the H⁺ and OH⁻ ions in the electrolyte.

Explanation:

The H^+ and OH^- ions in the electrolyte come from the auto-ionization of water:

$$H_2O(l) + H_2O(l) \rightleftharpoons H_3O^+(aq) + OH^-(aq).$$

If acid molecules are dissolved in water, the additional source of H_3O^+ ions come from the dissociated acid molecules. If a base is dissolved instead, then additional OH^- ions come from the alkali.

(iii) Explain why the electrolysis of concentrated aqueous sodium chloride liberates chlorine rather than oxygen at the anode.

Explanation:

At the anode, although OH^- is preferentially discharged, it is unlikely to happen as the concentration of OH^- is extremely low. Thus, it is the high concentration of Cl^- ions that undergoes discharge.

 Q So, does this mean that the concentrations of the species would affect which species is discharged at the electrode?

A: Yes, you are right! In a nutshell, the *concentration* of a species in electrolysis does affect whether it gets discharged or not!

Do you know?

— Brine, which is simply concentrated sodium chloride solution, is an important starting material for the production of *hydrogen*, *chlorine* gases, and *sodium hydroxide*. The reaction of chlorine gas and sodium hydroxide is important for the production of sodium chlorate(I) (NaClO), a powerful oxidizing agent, which acts as the active ingredient in *bleaching agent*:

$$2NaOH(aq) + Cl_2(aq) \rightarrow NaOCl(aq) + NaCl(aq) + H_2O(l).$$

The ions that are present during electrolysis would be similar to those for the dilute solution!

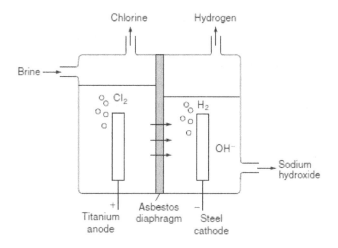

Electrolysis of brine

At the positively charged anode:

$$2Cl^-(aq) \rightarrow Cl_2(g) + 2e^- \quad \text{Oxidation.}$$

At the negatively charged cathode:

$$2H^+(aq) + 2e^- \rightarrow H_2(g) \quad \text{Reduction.}$$

Overall ionic equation:

$$2Cl^-(aq) + 2H^+(aq) \rightarrow Cl_2(g) + H_2(g).$$

Q   Is the above process used to manufacture chlorine gas?

A: Yup! The chlorine gas that is produced can be used to disinfect water.

> (iv) Explain why the electrolysis of concentrated aqueous sodium chloride liberates hydrogen rather than sodium at the cathode.

Explanation:

It is more difficult to discharge Na^+ than H^+ because from the reactivity series, sodium metal is above hydrogen. This means that sodium is more likely to undergo oxidation than hydrogen. Hence, Na^+ is less likely to be reduced than H^+.

 Q   But even if the Na^+ is reduced to $Na(s)$, would $Na(s)$ still be able to react with the water that is present in the electrolyte?

A: Yes! Na metal would react with the water to give hydrogen gas. So, it is back to square one.

> (v) Explain why the concentration of sodium hydroxide in the electrolyte increases during the electrolysis of concentrated aqueous sodium chloride.

Explanation:

Both H^+ and OH^- ions come from the auto-ionization of water. Thus, as H^+ is reduced to H_2, the concentration of OH^- ions would increase. In addition, since Cl^- ions are oxidized to Cl_2 while leaving Na^+ ions behind, the concentration of sodium hydroxide in the electrolyte increases during the electrolysis.

> (vi) Explain why the electrolysis of dilute aqueous copper(II) sulfate liberates copper metal rather than hydrogen at the cathode.

Explanation:

It is more difficult to discharge H^+ than Cu^{2+} because from the reactivity series, copper metal is below hydrogen. This means that hydrogen is more likely to undergo oxidation than copper. Hence, H^+ is less likely to be reduced than Cu^{2+}.

Do you know?

— During the electrolysis of aqueous $CuSO_4$ using inert electrodes, the ions that are present in the solution are: Cu^{2+}, SO_4^{2-}, H^+, and OH^-. Thus,

At the positively charged anode:
$$4OH^-(aq) \rightarrow 2H_2O(l) + O_2(g) + 4e^- \text{Oxidation.}$$
At the negatively charged cathode:
$$Cu^{2+}(aq) + 2e^- \rightarrow Cu(s) \text{Reduction.}$$
Overall ionic equation:
$$4OH^-(aq) + 2Cu^{2+}(aq) \rightarrow 2H_2O(l) + O_2(g) + 2Cu(s).$$
The OH^- and Cu^{2+} ions selectively get discharged at the anode and cathode, respectively!

— If aqueous $CuSO_4$ is electrolyzed using copper electrodes instead, the ions present are still: Cu^{2+}, SO_4^{2-}, H^+, and OH^-. But,

At the positively charged anode:
$$Cu(s) \rightarrow Cu^{2+}(aq) + 2e^- \text{Oxidation.}$$
At the negatively charged cathode:
$$Cu^{2+}(aq) + 2e^- \rightarrow Cu(s) \text{Reduction.}$$
Overall ionic equation:
$$Cu^{2+}(aq) + Cu(s) \rightarrow Cu(s) + Cu^{2+}(aq).$$
At the anode, it is the copper metal that undergoes oxidation and not $OH^-(aq)$. This is the basis for the purification of copper:

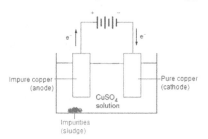

(*Continued*)

(Continued)

The impure copper is made the anode while a piece of pure copper acts as the cathode. As time passes, the anode dissolves, leaving the impurities behind while the pure copper grows in size.

In a nutshell, the *types of electrode* used in electrolysis can affect the outcome of the process!

Q Why is it important to purify copper?

A: Pure copper has lower resistance; this would help to minimize the loss of electrical energy through heat.

Q If we compare electroplating to the purification of copper, these two electrolytic processes are very similar. Am I right?

A: It is great that you managed to identify the similarities between the two processes. Electroplating is primarily used to coat a thin layer of material (of desirable properties) onto another material (which lacks the desired property). For instance, a popular use is in the electroplating of jewellery. Inexpensive jewellery is often coated with a thin layer of a precious metal such as silver or gold.

The piece of metal that is to be coated with silver, is made the cathode, which is very similar to the making of a piece of pure copper as the cathode during the purification process. This cathode is than placed into an electrolytic solution that contains the ions of the coating material, i.e., Ag^+ ions.

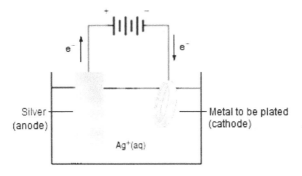

Silver (anode) — — $Ag^+(aq)$ — — Metal to be plated (cathode)

The coating material, i.e., Ag(s) is made the anode, which is very similar to the making of a piece of impure copper as the anode during the purification process.

When an electric current is passed, the Ag^+ ions in the solution will migrate to the cathode and undergo reduction to form Ag(s) on the surface of the cathode.

At the positively charged anode:

$$Ag(s) \rightarrow Ag^+(aq) + e^- \quad \text{Oxidation.}$$

At the negatively charged cathode:

$$Ag^+(aq) + e^- \rightarrow Ag(s) \quad \text{Reduction.}$$

Overall ionic equation:

$$Ag^+(aq) + Ag(s) \rightarrow Ag(s) + Ag^+(aq).$$

(vii) During the electrolysis of dilute aqueous copper(II) sulfate using inert electrodes, the blue coloration of the aqueous copper(II) sulfate fades and the pH of the solution decreases after some time. Explain why these changes take place.

Explanation:

The blue coloration of the aqueous copper(II) sulfate fades because the Cu^{2+}(aq) ions, which give rise to the blue coloration, have been reduced to copper metal at the cathode.

The pH of the solution decreases after some time is due to the oxidation of OH^-(aq) at the anode, as follows:

$$4OH^-(aq) \rightarrow 2H_2O(l) + O_2(g) + 4e^-.$$

As both OH^-(aq) and H^+(aq) ions come from the auto-ionization of water, when the concentration of OH^-(aq) decreases, the concentration of H^+(aq) increases relatively. Thus, the pH of the solution thus decreases.

(viii) State some possible uses for the hydrogen gas produced.

Explanation:

Hydrogen can be used (i) as a fuel in the fuel cell; (ii) to hydrogenate fats using nickel catalyst; and (iii) in the synthesis of ammonia from nitrogen in the Haber Process.

> (ix) Give the dot-and-cross diagram of potassium chloride.

Explanation:

$$\left[K \right]^{+} \left[:\overset{..}{\underset{..}{C}} l_{x} \right]^{-}$$

> (x) State the types of bonding in molten potassium chloride.

Explanation:

In molten potassium chloride, the oppositely charged ions are attracted to each other by ionic bonds.

 Q So, we did not actually break the ionic bonds in an ionic compound when we melted it?

A: Nope! You just weakened them.

> (b) Aqueous copper(II) sulfate was electrolyzed using copper electrodes. The copper anode lost mass as copper(II) ions were formed, and the copper cathode gained mass as copper atoms were formed.
>
> (i) State one industrial application of this electrolysis.

Explanation:

The impure copper can be made the anode while the pure copper can be made the cathode. In such a way, we can get purified copper. Thus, this method is used to purify copper in the industry.

> (ii) Explain the likely factors that would affect the rate of the deposition of copper at the cathode.

Explanation:

How fast the copper can be deposited would depend on the quantity of electrons that flow through the system per unit time. This is determined by the magnitude of the current.

In addition, the concentration of the copper(II) sulfate solution would also affect the rate of deposition. The higher the concentration of the solution, the higher the electrical conductivity; hence, the faster the rate of deposition of the copper.

The temperature would also affect the rate of deposition. The higher the temperature, the higher would be the rate of collision of the particles with the surface of the electrode; hence, the faster the rate of deposition.

> (iii) Explain why copper is such a good conductor of electricity.

Explanation:

Copper has delocalized valence electrons which can act as mobile charge carriers when a potential difference is applied across the metal. This accounts for its good electrical conductivity.

 Q Since a metal and an electrolyte can conduct electricity, are there any differences between them?

A: The charge carriers in a metal are delocalized electrons while those in an electrolyte are free ions. In addition, a piece of metal remains unchanged after electrical conduction but for an electrolyte, it undergoes discharged or it "decomposes."

2. An aqueous solution of barium hydroxide is electrolyzed between carbon electrodes.

    (a) What gas would you expect to be produced at the anode? Give the species that produces the gas.

Explanation:

The species that are present in aqueous barium hydroxide are: Ba^{2+}, H^+, OH^-, and H_2O. At the positively charged anode, oxidation takes place. Thus, the species that would likely be attracted to the anode is the OH^- ion and it will be oxidized to form oxygen gas:

$$4OH^-(aq) \rightarrow 2H_2O(l) + O_2(g) + 4e^-.$$

    (b) How would you test for the gas evolved at the anode?

Explanation:

Oxygen gas can be tested by a glowing splint. The gas would relight a glowing splint.

    (c) Give the half-equations for the reaction at the anode and cathode.

Explanation:

Anode (oxidation): $4OH^-(aq) \rightarrow 2H_2O(l) + O_2(g) + 4e^-.$

At the cathode, reduction can only occur for H^+, forming hydrogen gas.

Cathode (reduction): $2H^+(aq) + 2e^- \rightarrow H_2(g).$

    (d) Give the molar ratio of the products that are formed at the anode to the cathode.

Explanation:

A balanced redox equation is:

$$4OH^-(aq) + 4H^+(aq) \rightarrow 2H_2O(l) + O_2(g) + 2H_2(g).$$

Hence, the molar ratio of O_2: H_2 is 1:2.

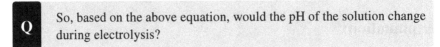

So, based on the above equation, would the pH of the solution change during electrolysis?

A: Since equal amounts of $OH^-(aq)$ and $H^+(aq)$ are "consumed" during the electrolytic process, the pH is not going to change!

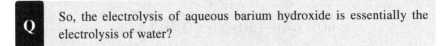

So, the electrolysis of aqueous barium hydroxide is essentially the electrolysis of water?

A: You are right! Essentially, it is just this reaction: $2H_2O(l) \rightarrow 2H_2(g) + O_2(g).$

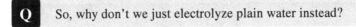

So, why don't we just electrolyze plain water instead?

A: It is difficult to electrolyze plain water to get hydrogen and oxygen as the electrical conductivity of plain water is very low. This is due to the minute amount of ions that is generated from the auto-ionization of water. You can simply treat the presence of barium hydroxide as just to increase the electrical conductivity of the solution!

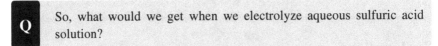

So, what would we get when we electrolyze aqueous sulfuric acid solution?

A: Simple! The ions that are present in aqueous sulfuric acid are: SO_4^{2-}, H^+, and OH^-. Since SO_4^{2-} ions are difficult to be discharged, we have the following reactions occurring at the electrodes:

At the positively charged anode:

$$4OH^-(aq) \rightarrow 2H_2O(l) + O_2(g) + 4e^- \text{   Oxidation.}$$

At the negatively charged cathode:

$$2H^+(aq) + 2e^- \rightarrow H_2(g) \text{   Reduction}$$

Overall ionic equation:

$$4OH^-(aq) + 4H^+ \rightarrow 2H_2O(l) + O_2(g) + 2H_2(g).$$

Thus, the electrolytic process is simply the electrolysis of plain water!

Do you know?

— The electrolysis of dilute aqueous sodium chloride solution is also equivalent to the electrolysis of plain water. The ions present are: Na^+, Cl^-, H^+, and OH^-. Both Na^+ and Cl^- ions would not undergo discharge as they are less likely to do so. Thus, when we electrolyze this solution using an inert electrode of platinum or carbon, we would have the following reactions happening at the cathode and anode:

At the positively charged anode:

$$4OH^-(aq) \rightarrow 2H_2O(l) + O_2(g) + 4e^- \quad \text{Oxidation.}$$

At the negatively charged cathode:

$$2H^+(aq) + 2e^- \rightarrow H_2(g) \quad \text{Reduction.}$$

Overall ionic equation:

$$4OH^-(aq) + 4H^+ \rightarrow 2H_2O(l) + O_2(g) + 2H_2(g).$$

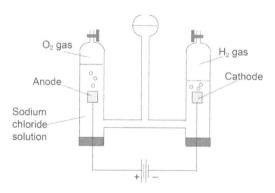

(e) It is observed that, in fact, during the electrolysis, the mass of the anode decreases and a white precipitate forms around it.

   (i) Suggest an explanation for these observations.

270 Understanding Basic Chemistry Through Problem Solving

Explanation:

As oxygen gas is evolved at the anode, it can react with the carbon electrode to give carbon monoxide and carbon dioxide gases. This accounts for the decreases in the mass of the anode. As the carbon dioxide dissolves in the alkaline solution, the CO_2 undergoes the following reaction:

$$CO_2(g) + 2OH^-(aq) \rightarrow CO_3{}^{2-}(aq) + H_2O(l).$$

The $CO_3{}^{2-}(aq)$ ions form the insoluble $BaCO_3$ precipitate, which is white in color.

(ii) Give an equation that can be used to explain the decrease in mass of the anode.

Explanation:

$$C(s) + O_2(g) \rightarrow CO_2(g).$$

(iii) Identify the white precipitate and give the equation for its formation.

Explanation:

The white precipitate is $BaCO_3$.

(iv) Explain why the white precipitate has a high melting point.

Explanation:

Barium carbonate is an ionic compound with strong ionic bonds between the oppositely charged ions. This accounts for its high melting point.

If barium carbonate has a high melting point due to the strong ionic bonds, then why does it decompose upon heating?

A: The melting and decomposition processes are two separate issues. Barium carbonate decomposes to barium oxide (BaO) and carbon dioxide due to the high charge density of the Ba^{2+} ion. The high charge density "pulls" the electron cloud of the CO_3^{2-} ion toward itself. This in turn weakens the intramolecular covalent bonds within the CO_3^{2-} ion, causing it to be easily broken up upon heating. Thus, when barium carbonate decomposes, it has not melted!

(3) Sodium chloride has a melting point of 801°C. Electrolysis of molten sodium chloride was carried out under an inert atmosphere of argon gas with platinum as the electrodes.

   (a) Explain why sodium chloride has such a high melting point.

Explanation:

Sodium chloride is an ionic compound with strong ionic bonds between the oppositely charged ions. This accounts for its high melting point.

   (b) Explain why molten sodium chloride and not solid sodium chloride was used for electrolysis.

Explanation:

In solid sodium chloride, the ions are rigidly held in the solid lattice structure. Hence, they cannot act as mobile charge carriers. In contrast, in the molten state, the ions are mobile and thus can act as charge carriers.

   (c) Give the dot-and-cross diagram of sodium chloride.

Explanation:

$$\left[\text{Na}\right]^{+} \left[:\overset{\displaystyle\cdot\cdot}{\underset{\displaystyle\cdot\cdot}{\text{Cl}}}:\right]^{-}$$

(d) Give the half-equations for the reactions that occurred at the anode and cathode. Hence, give the overall reaction equation with state symbols.

Explanation:

At the positively charged anode:

$$2Cl^- \rightarrow Cl_2 + 2e^- \quad \text{Oxidation.}$$

At the negatively charged cathode:

$$Na^+ + e^- \rightarrow Na \quad \text{Reduction.}$$

Overall reaction equation:

$$2Cl^-(l) + 2Na^+(l) \rightarrow Cl_2(g) + 2Na(l).$$

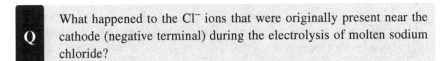

Q Why is the state symbol for the sodium "(l)"? Shouldn't it be "(s)"?

A: Under the high temperature, the sodium would have already melted.

Q What happened to the Cl^- ions that were originally present near the cathode (negative terminal) during the electrolysis of molten sodium chloride?

A: The Cl^- near the negative terminal would be repelled by the negatively charged electrode, and likewise for the Na^+ ions that were originally near the positive terminal (anode). So, the movement of charges in the electrolyte constitutes an electrical current. In addition, such a movement of charges also ensures that there is *electrical neutrality* in the electrolyte.

Q Does electrolysis help us verify the existence of ions?

A: Certainly! For instance, the ions in a solid ionic compound are not mobile but when an electrical potential is applied across molten ionic compound, it can actually conduct electricity. This is a demonstration of the presence of ions!

(e) Describe a test you would carry out to identify the product obtained at the anode. Give the reaction equation for the positive test.

Explanation:

A piece of filter paper strip soaked with starch and potassium iodide solutions can be used to test for the presence of chlorine. If the starch paper turns blue-black, then the gas is chlorine.

$$Cl_2(g) + 2I^-(aq) \rightarrow 2Cl^-(aq) + I_2(aq).$$

The chlorine oxidizes the iodide ions to form iodine, which forms a blue-black coloration with the starch.

Do you know?

— Group 17 elements are strong oxidizing agents. As a result of their high electronegativity, they tend to accept electrons readily from other substances and become reduced. The product is the halide ions, which carry a charge of −1 and has a stable octet configuration in the valence shell.

$$X_2 + 2e^- \rightarrow 2X^-, \quad X = F, Cl, Br, or I.$$

Since electronegativity *decreases* down the group, it is expected that the *ease* in accepting electrons also decreases. In other words, the oxidizing power of the Group 17 elements decreases down the group with fluorine being the strongest and iodine the weakest oxidizing agent in the group.

— One important reaction to show the decrease in oxidizing power down Group 17 is through the *displacement* reaction. In a displacement reaction, the halogen which is a stronger oxidizing agent will be able to oxidize the halide ion of the weaker oxidizing agent that is below itself in the group. The halide ion is said to be displaced from the solution.

(Continued)

(Continued)

— We can make use of this type of reaction to identify the presence of Br^- and I^- ions in a solution. Simply add aqueous Cl_2 followed by an organic solvent. Upon shaking the mixture, the presence of Br^- or I^- can be confirmed as Br_2 and I_2 provides distinct coloration in the organic layer.

$$Cl_2(aq) + 2Br^-(aq) \rightarrow 2Cl^-(aq) + Br_2(aq); \text{ or}$$
$$Cl_2(aq) + 2I^-(aq) \rightarrow 2Cl^-(aq) + I_2(aq).$$

If we try to react I_2 with Cl^-, no reaction will occur. I^- and Cl_2 are not produced!

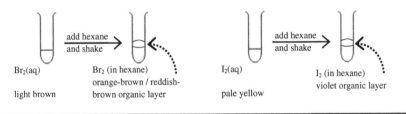

$Br_2(aq)$ → add hexane and shake → Br_2 (in hexane)
light brown   orange-brown / reddish-brown organic layer

$I_2(aq)$ → add hexane and shake → I_2 (in hexane)
pale yellow   violet organic layer

(f) State one problem you would expect when carrying out this experiment in the laboratory.

Explanation:

Chlorine is a very pungent gas that would irritate the eyes.

(g) Give the name of the reaction when the product at the cathode reacts with water. Hence, give an overall equation for this reaction.

Explanation:

$$2Na(s) + 2H_2O(l) \rightarrow 2NaOH(aq) + H_2(g).$$

This is a redox reaction.

Q Can we call it an acid–metal reaction?

A: It is in fact an acid–metal reaction as the water is behaving as an acid, but the expected answer is redox reaction. So, it is better not to.

> (h) Give the electronic configuration of the atom of the product produced at the cathode.

Explanation:

The electronic configuration of Na is 2.8.1.

> (i) Explain why the product at the cathode is highly reactive.

Explanation:

By losing an electron, the Na atom becomes the Na^+ ion, which has an electronic configuration of 2.8. It thus has a stable octet configuration. This explains why the sodium is highly reactive.

But you have said before that the octet rule is not the rationale why a sodium atom would prefer to lose an electron. So, why did you use it as an explanation here?

A: Well, what should you do? This is the requirement for the examination, you still need to answer according to what the examiner expects from you. As long as you know the actual reasoning, it is great!

> (j) Identify the polarity of these two electrodes.

Explanation:

Since Na^+ is attracted to the cathode to undergo reduction, the cathode is negatively charged. As the Cl^- is attracted to the anode to undergo oxidation, the anode is positively charged.

(k) Explain why platinum is used as the electrodes.

Explanation:

Platinum is an inert metal that does not react with the electrolytic products.

(l) Why was the electrolysis carried out under an inert atmosphere of argon.

Explanation:

Argon is an inert gas that does not react with the electrolytic products.

(m) What would be the result you would obtain if concentrated aqueous sodium chloride was used instead of molten sodium chloride?

Explanation:

If concentrated aqueous sodium chloride was used instead of molten sodium chloride, then the following reactions would take place at the anode and cathode:

At the positively charged anode:
$$2Cl^-(aq) \rightarrow Cl_2(g) + 2e^- \quad \text{Oxidation.}$$

At the negatively charged cathode:
$$2H^+(aq) + 2e^- \rightarrow H_2(g) \quad \text{Reduction.}$$

Overall ionic equation:
$$2Cl^-(aq) + 2H^+ \rightarrow Cl_2(g) + H_2(g).$$

(n) What would be the result you would obtain if dilute aqueous sodium chloride was used instead of molten sodium chloride?

Explanation:

If dilute aqueous sodium chloride was used instead of molten sodium chloride, then the following reactions would take place at the anode and cathode:

At the positively charged anode:

$$4OH^-(aq) \rightarrow 2H_2O(l) + O_2(g) + 4e^- \quad \text{Oxidation.}$$

At the negatively charged cathode:

$$2H^+(aq) + 2e^- \rightarrow H_2(g) \quad \text{Reduction.}$$

Overall ionic equation:

$$4OH^-(aq) + 4H^+(aq) \rightarrow 2H_2O(l) + O_2(g) + 2H_2(g).$$

(4) The extraction of aluminum from its ore is done by electrolyzing a mixture of aluminum oxide and cryolite ($NaAlF_4$) at 800°C using carbon electrodes.

(a) Give the dot-and-cross diagram of aluminum oxide.

Explanation:

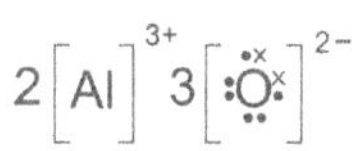

$$2\left[Al\right]^{3+} 3\left[\overset{\times}{\underset{\bullet\bullet}{:}}\overset{\bullet\bullet}{O}\overset{\times}{:}\right]^{2-}$$

Do you know?

— As aluminium is above carbon in the reactivity series, it is extracted through electrolysis. There are two important stages involved in the extraction process of pure aluminum:

(Continued)

(*Continued*)

Stage 1: *Concentrating the ore*

The mineral must first be separated from most of the worthless compounds in the bauxite ore, usually by physical methods. But for aluminum, bauxite is reacted with hot concentrated sodium hydroxide to isolate the aluminum oxide:

$$Al_2O_3 + 2NaOH + 3H_2O \rightarrow 2NaAl(OH)_4.$$

The hot solution is then cooled to crystallize out the aluminum(III) hydroxide:

$$NaAl(OH)_4 \rightarrow Al(OH)_3 + NaOH.$$

Then, the aluminum(III) oxide is obtained by dehydrating the aluminum(III) hydroxide:

$$2Al(OH)_3 \rightarrow Al_2O_3 + 3H_2O.$$

Stage 2: *Extracting the crude metal from the ore*

The aluminum oxide is electrolyzed by dissolving it in molten cryolite, Na_3AlF_6, at about $1000°C$. This helps to save electrical energy as aluminum oxide melts at about $2000°C$. The following show the electrolytic set-up:

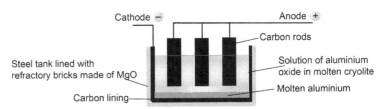

Simplistically, the following reactions are thought to happen:

Anode: $2O^{2-} \rightarrow O_2 + 4e^-$
Cathode: $Al^{3+} + 3e^- \rightarrow Al$
Overall ionic equation: $4Al^{3+}(l) + 6O^{2-}(l) \rightarrow 4Al(l) + 3O_2(g).$

As the temperature of the electrolytic bath is high, the oxygen reacts with the carbon electrodes to form carbon monoxide and carbon dioxide gases. As a result, the electrodes need constant replacement. The molten aluminum that is formed at the anode can simply be drained off from the bottom of the bath.

(b) Why is a mixture of aluminum oxide and cryolite used in the above instead of pure aluminum oxide?

Explanation:

Aluminum oxide melts at about 2000°C. With the addition of cryolite, the melting point is lowered to about 1000°C. This helps to save electrical energy.

(c) Write the ionic equations for the reactions that occurred at the anode and cathode.

Explanation:

Anode: $2O^{2-} \rightarrow O_2 + 4e^-$.
Cathode: $Al^{3+} + 3e^- \rightarrow Al$.

(d) After some time, the carbon electrode decreased in mass and was needed to be replaced. Explain.

Explanation:

As the temperature of the electrolytic bath is high, the oxygen reacts with the carbon electrodes to form carbon monoxide and carbon dioxide. As a result, the electrodes need constant replacement.

(e) Write an equation to help explain the decrease in mass of the anode.

Explanation:

$$C(s) + O_2(g) \rightarrow CO_2(g).$$

(f) Given the trouble of regular replacement of a graphite anode, would you suggest replacing the graphite anode with a steel anode? Explain your reason.

Explanation:

I would not suggest replacing the graphite anode with a steel anode. This is because although the melting point of steel is higher than the operating temperature of the electrolytic process, the structure of steel may actually weaken during the high operating temperature.

Q Can I give "yes" as an answer instead?

A: Well, you can. Then all you need to do is to give the reason that steel has a high melting point, thus there is no need for regular replacement. But if steel can really replace carbon as the anode electrode, then why did your teacher explain to you that carbon electrodes are still used in the electrolysis of aluminum oxide?

(g) Aluminum is commonly used to make saucepans although steel is stronger and cheaper. Give a reason.

Explanation:

Aluminum is relatively more resistant to corrosion than steel because the layer of aluminum oxide coated on the surface is impervious to further chemical attack by air and water. In addition, aluminum is lighter than steel. This makes the saucepans lighter, and hence, easier to handle.

Do you know?

— Although the layer of aluminum oxide can be formed naturally in air, the layer is not thick enough. The layer of aluminum oxide on the surface of the saucepan can be increased via *anodization*, i.e., by making the aluminum objects as the anode during the electrolytic process. The layer of aluminum oxide is of substantial commercial and technological importance for the prevention of corrosion for automobile and aerospace structures and also for electrical insulation.

— Anodization of aluminum is usually carried out in a sulfuric acid electrolyte. During the electrolytic process of anodization, oxygen gas is discharged at the anode, which would react with the unoxidized aluminum metal to form a thick aluminum(III) oxide layer. The freshly formed film can then further be dyed to give color-anodized aluminum. The cathode can be graphite, stainless steel, or other electrical conductors that are inert in the anodizing bath.

At the positively charged anode:

$$4OH^-(aq) \rightarrow 2H_2O(l) + O_2(g) + 4e^-$$
$$4Al(s) + 3O_2(g) \rightarrow 2Al_2O_3(s).$$

At the negatively charged cathode:

$$2H^+(aq) + 2e^- \rightarrow H_2(g).$$

 Q If the electrolyte is sulfuric acid, wouldn't it react with the amphoteric Al_2O_3?

A: It would not react because the H^+ would be repelled from the anode, which is positively charged!

(h) Calculate the maximum possible mass of aluminum that can be extracted from 10 kg of aluminum oxide.

Explanation:

Molar mass of $Al_2O_3 = 2(27.0) + 3(16.0) = 102.0\,g\,mol^{-1}$.

Amount of Al_2O_3 in $10\,kg = \dfrac{10,000}{102.0} = 98.0\,mol$.

Amount of Al that can be extracted $= 2 \times 98.0 = 196.0\,mol$.

Maximum possible mass of aluminum that can be extracted $= 196.0 \times 27.0 = 5294\,g$.

> (i) Aluminum is extracted from aluminum oxide. A major use of aluminum is in the manufacture of drink cans. A typical drink can contains 9.0 g of aluminum. Calculate the mass of aluminum oxide required to make a drink can.

Explanation:

Amount of Al in 9.0 g $= \dfrac{9.0}{27.0} = 0.333\,mol$.

Amount of Al_2O_3 needed $= \dfrac{1}{2} \times$ Amount of Al in 9.0 g $= \dfrac{1}{2} \times 0.333$
$$= 0.167\ mol.$$

Mass of Al_2O_3 needed $= 0.167 \times 102.0 = 17.0\,g$.

> (j) Suggest why manufacturers of drink cans are always trying to find ways of reducing the mass of aluminum required for a typical drink can.

Explanation:

By reducing the mass of aluminum that is required for a typical drink can, less aluminum oxide is needed. More money can be saved as smaller amount of energy would be used to extract a smaller amount of aluminum from the oxide.

(k) Explain why the recycling of aluminum is economically viable.

Explanation:

The costs involved in the extraction, especially the amount of electrical current needed, are high. Usually for metals that are obtained through electrolysis, it is not cheap. Thus, recycling of metals obtained via electrolytic means is viable.

CHAPTER 12

THE PERIODIC TABLE

1. (a) Explain what do you understand by the terms (i) *atom*, (ii) *atomic number*, (iii) *relative atomic mass*, (iv) *empirical formula*, and (v) *molecular formula*.

Explanation:

 (i) An atom is the smallest building block of matter. It consists of electrons, protons, and neutrons.

 (ii) The atomic number indicates the number of protons present in an atom of an element. Different elements consist of atoms with different proton numbers. In a neutral atom, the atomic number is also equal to the number of electrons present.

(iii) The relative atomic mass is the average mass of one atom of the element relative to 1/12 of the mass of a carbon-12 atom. The relative atomic number is a weighted average of the masses of the different types of isotopes that may be present for a particular element.

 (iv) The empirical formula (EF) of a compound is a representation of the simplest whole number ratio of the different atoms that made up the compound.

 (v) The molecular formula (MF) of a compound is a representation of the actual whole number ratio of the different atoms that made up the compound. Thus, MF = n × EF, i.e., an empirical formula is a fraction of the molecular formula.

Q Why don't we use the number of neutrons or number of electrons to characterize an element?

A: We have seen that an element can be made up of isotopes with the same number of protons but different number of neutrons. So, using the number of neutrons to characterize an element is not very good — which isotope would you prefer? How do you decide? Now, different species can have the same number of electrons but different number of protons or neutrons, so again, using the number of electrons to characterize an element cannot be as unique as using the number of protons to characterize it. This is because the number of protons for an element is unique to that element and protons cannot be lost through normal chemical means.

(b) Discuss briefly, giving examples, the variations of chemical and physical properties to be found in (i) a period and (ii) a group of the Periodic Table.

Explanation:

(i) As we go from left to right of a period, the elements shift from *metals* to *non-metals*. The atoms in a metal are bonded together by metallic bonds (e.g., Na, Mg, and Al), while those of the non-metals are bonded by covalent bonds (e.g., Si, P, S, and Cl).

Metals tend to lose electrons to form positively charged cations (e.g., Na^+, Mg^{2+}, and Al^{3+}). In contrast, non-metals tend to gain electrons to form negatively charged anions (e.g., P^{3-}, S^{2-}, and Cl^-). Thus, when a metal and non-metal react, they would *likely* form ionic compounds (e.g., NaCl), whereas when two non-metals react, they would likely form covalent compounds (e.g., CO_2).

(ii) The metals in Group 1 have low melting and boiling points due to the weak metallic bonds. As we go down Group 1, the metallic bond strength decreases because of an increase in the sizes of the atoms. But for Group 17 elements, the melting point increases down the group due to an increase in the strength of the intermolecular forces.

The reducing power of elements in any particular group increases down the group. For example, for Group 1 elements, possessing only one valence electron which is easily lost, their reducing power increases down the group. This would also mean that the oxidizing power would decrease down a group, especially for the non-metals (Groups 15, 16 and 17).

Q Is there any significance to the period number?

A: Yes, the *period number*, assigned to the *horizontal* rows, indicates one important feature for an atom of a specific element — the number of electronic shells the atom has. For example:

Period number	Element	Proton number	Number of electrons	Electronic configuration
2	Lithium, Li	3	3	2.1
2	Beryllium, Be	4	4	2.2
2	Boron, B	5	5	2.3
2	Carbon, C	6	6	2.4
2	Neon, Ne	10	10	2.8
3	Sodium, Na	11	11	2.8.1
3	Chlorine, Cl	17	17	2.8.7
3	Argon, Ar	18	18	2.8.8

Q How about the group number?

A: The *group number*, assigned to the *vertical* columns, represents the number of valence electrons in the valence electronic shell after subtracting 10 units from the group number for elements that come from Groups 13 to 18. For examples:

Element	Proton number	Number of electrons	Electronic configuration	Group number
Hydrogen, H*	1	1	1	1
Helium, He*	2	2	2	18
Lithium, Li	3	3	2.1	1
Beryllium, Be	4	4	2.2	2

(Continued)

		(Continued)		
Element	Proton number	Number of electrons	Electronic configuration	Group number
Boron, B	5	5	2.3	13
Carbon, C	6	6	2.4	14
Nitrogen, N	7	7	2.5	15
Oxygen, O	8	8	2.6	16
Fluorine, F	9	9	2.7	17
Neon, Ne	10	10	2.8	18

Hydrogen (H) and helium (He) are exceptionally placed in Groups 1 and 18, respectively. Hydrogen like other Group 1 elements, has only one oxidation state of +1, while helium is chemically similar to the elements in Group 18.

 Q What causes a metal atom to lose electrons more readily than a non-metal atom across a period?

A: The main characteristic of a metal is that it tends to *lose* electrons (greater electropositivity), while for a non-metal, it tends to *gain* electrons (greater electronegativity). This is because of the change in electronegativity values across a period. The increase in *electronegativity* (or decrease in electropositivity) across a period from left to right, is actually due to an increase in the net electrostatic attractive force acting on the valence electrons. With stronger electrostatic forces acting on the valence electrons, more energy is required to remove the valence electrons for the non-metals. Hence, when a non-metal "encounters" a metal, the non-metal atom would gain electrons, while the metal atom would lose electrons. Similarly, when two non-metal atoms "meet," sharing of electrons is the best "solution."

 Q As we go down a group, the size of an atom increases, hence the distance of the valence shell from the nucleus also increases. Therefore, the valence electrons are less strongly attracted by the nucleus. Does this mean that the electronegativity of the atom would also decrease down a group?

A: Absolutely right! Down a group, there is an increase in metallic character and an increase in electropositivity or a decrease in electronegativity. This also accounts for an increase in reducing power (losing electrons and be oxidized) or a decrease in oxidizing power (gaining electrons and be reduced) down the group!

Q Wow! I also noticed that the ease to remove an electron decreases across a period BUT it increases down a group. Am I right?

A: That is sharp of you. Yes, it is actually because of these two trends that elements which are diagonally across each other in the Periodic Table have similar chemical properties. This is known as a *diagonal relationship*. For more details, you can refer to *Understanding Advanced Physical Inorganic Chemistry* by J. Tan and K.S. Chan.

Do you know?

— The table below summarizes the physical and chemical properties of Period 3 elements and their oxides across the period:

Group	1	2	13	14	15	16	17	18
Name	Sodium	Magnesium	Aluminum	Silicon	Phosphorous	Sulfur	Chlorine	Argon
Symbol	Na	Mg	Al	Si	P	S	Cl	Ar
Structure of element	Na	Mg	Al	Si	P_4	S_8	Cl_2	Ar
Property	Metal			Metalloid	Non-metal			
Bonding in element	Metallic bond			Covalent bond				Mono-atomic
Structure of oxide	Giant ionic			Giant covalent	Simple molecular			
Bonding in oxide	Ionic bond			Covalent bond	Covalent bond *within* the molecule and IMF* *between* molecules			
Nature of oxide	Basic		Amphoteric	Acidic				
Chemical formula of oxide	Na_2O	MgO	Al_2O_3	SiO_2	P_4O_{10}	SO_3		
Oxidation state	+1	+2	+3	+4	+5	+6		
Solubility of oxide	Soluble	Insoluble	Insoluble	Insoluble	Soluble	Soluble		

*Intermolecular forces.

(*Continued*)

(*Continued*)

Take note that the maximum oxidation state that the element can assume corresponds to the number of valence electrons it has, which is the (Group number − 10) for elements that come from Groups 13 to 18! Similar trends have also been observed for the Period 3 chlorides below:

Group	1	2	13	14	15	16	17	18
Name	Sodium	Magnesium	Aluminum	Silicon	Phosphorous	Sulfur	Chlorine	Argon
Symbol	Na	Mg	Al	Si	P	S	Cl	Ar
Structure of chloride	Giant ionic			Ionic in solid but simple molecular when melted	Simple molecular			
Bonding in chloride	Ionic bond		Covalent bond *within* the molecule and IMF *between* molecules					
Nature of chloride	Neutral	Weakly acidic	Acidic					
Chemical formula of chloride	NaCl	$MgCl_2$	$AlCl_3$	$SiCl_4$	PCl_5	SCl_6	Cl_2	
Oxidation state	+1	+2	+3	+4	+5	+6		
Solubility of chloride	Soluble	Soluble	Soluble	Soluble	Soluble	Soluble		

2. The elements sodium, magnesium, aluminium, silicon, phosphorous, sulfur, chlorine, and argon are the elements in Period 3. For each of the elements, state:

(a) the electronic structure;

Explanation:

Element	Number of electrons	Electronic configuration
Sodium, Na	11	2.8.1
Magnesium, Mg	12	2.8.2
Aluminum, Al	13	2.8.3
Silicon, Si	14	2.8.4
Phosphorous, P	15	2.8.5
Sulfur, S	16	2.8.6
Chlorine, Cl	17	2.8.7
Argon, Ar	18	2.8.8

(b) the structure and bonding in the element;

Explanation:

Group	1	2	13	14	15	16	17	18
Name	Sodium	Magnesium	Aluminum	Silicon	Phosphorous	Sulfur	Chlorine	Argon
	Na	Mg	Al	Si	P_4	S_8	Cl_2	Ar
Structure of element	Metallic lattice	Metallic lattice	Metallic lattice	Macro-molecular/ Giant covalent	Simple molecular	Simple molecular	Simple molecular	Mono-atomic
Bonding in element	Metallic bond			Covalent bond				Mono-atomic

(c) the formula of its hydride and the reaction, if any, between the hydride and water;

Explanation:

Name	Formula of hydride	Reaction with water
Na	NaH	$NaH + H_2O \rightarrow NaOH + H_2$
Mg	MgH_2	$MgH_2 + 2H_2O \rightarrow Mg(OH)_2 + 2H_2$
Al	AlH_3	$AlH_3 + 3H_2O \rightarrow Al(OH)_3 + 3H_2$
Si	SiH_4	–
P	PH_3	$PH_3 + H_2O \rightleftharpoons PH_4^+ + OH^-$
S	H_2S	$H_2S + H_2O \rightleftharpoons HS^- + H_3O^+$
Cl	HCl	$HCl + H_2O \rightarrow Cl^- + H_3O^+$
Ar	–	–

 Q Why is hydrogen gas evolved for the reactions of NaH, MgH_2, and AlH_3, with water?

A: NaH, MgH_2, and AlH_3 are all ionic compounds containing hydride, H^-, ions. So, naturally, when these hydrides react with water molecules, the electron-rich H^- ion and the electron-deficient hydrogen atom of the H–O–H molecule can undergo an acid–base reaction. So, you can imagine the H^- ion slowly pulling out a hydrogen atom from the H_2O molecule to form H_2 gas.

Q So, why doesn't SiH_4 react with water?

A: The Si atom is slightly less electronegative than the H atom. As a result, the H atoms of SiH_4 are only slightly electron rich, i.e., possessing δ- partial charge, which is not strong enough to react with the water molecule.

Q So, the reaction of PH_3 with water is very similar to NH_3. Is it because both P and N are from the same group?

A: Elements from the same group have similar chemical properties because they have the same number of valence electrons. Since NH_3 is basic, we can predict PH_3 to be also basic in nature.

Q So, as H_2S can protonate H_2O, does it mean that H_2S is a stronger acid than H_2O?

A: Yes, you are right! The reason is because the H–S bond is weaker than the H–O bond.

(d) the formula of an oxide and whether the oxide is acidic, basic, amphoteric, or neutral;

(e) the structure and bonding of the oxide;

Explanation:

Group	1	2	13	14	15	16	17	18
Name	Sodium	Magnesium	Aluminum	Silicon	Phosphorous	Sulfur	Chlorine	Argon
Structure of oxide	Giant ionic			Giant covalent	Simple molecular			
Bonding in oxide	Ionic bond			Covalent bond	Covalent bond *within* the molecule and IMF *between* molecules			
Nature of oxide	Basic		Amphoteric	Acidic				
Chemical formula of oxide	Na_2O	MgO	Al_2O_3	SiO_2	P_4O_{10}	SO_3		

 Q Why are Na_2O and MgO both basic in nature whereas Al_2O_3 is amphoteric?

A: Although Na_2O, MgO, and Al_2O_3 are all metal oxides, they are basic in nature because of the presence of the electron-rich O^{2-} ions, which can easily react with an acid. In addition, Al_2O_3 can react with a base, such as OH^-.

This is because of the presence of the Al^{3+} ion, which has a high charge density. The charge density of the positive cation is so great that it is strong enough to attract the OH^- ion, which has a charge opposite of the cation, to form another particle as follows:

$$Al_2O_3(aq) + 6HCl(aq) \rightarrow 2AlCl_3(aq) + 3H_2O(l)$$

$$Al_2O_3(s) + 2NaOH(aq) \rightarrow 2NaAlO_2(aq) + H_2O(l).$$
$$\text{sodium aluminate}$$

Q   Why are non-metal oxides acidic in nature?

A: If you look at the reaction of SO_2, a non-metal oxide, in water:

$$SO_2(g) + H_2O(l) \rightleftharpoons H_2SO_3(aq).$$

What do you notice? The O atom of H_2O molecule becomes bonded to the S atom of SO_2. How does it happen? This is because the S atom of SO_2 is highly electron deficient as it is bonded to highly electronegative O atoms. The electronegative O atoms withdraw electron density away from the S atom, making it electron deficient and hence easily attracted to the electron-rich atom of the H_2O molecule.

Q   So, is the above explanation also applicable to explain why non-metal chlorides, such as $SiCl_4$, are acidic in nature?

A: Yes! Look at the equation below:

$$SiCl_4 + 4H_2O \rightarrow Si(OH)_4 + 4HCl;$$

how are the O atoms from the H_2O molecules being bonded to the Si atom?

(f) the formula of a chloride and whether the chloride is acidic, basic, or neutral; and

(g) the structure and bonding of the chloride.

Explanation:

Group	1	2	13	14	15	16	17	18
Name	Sodium	Magnesium	Aluminum	Silicon	Phosphorous	Sulfur	Chlorine	Argon
Symbol	Na	Mg	Al	Si	P	S	Cl	Ar
Structure of chloride	Giant ionic		Ionic in solid but simple molecular when melted	Simple molecular				
Bonding in chloride	Ionic bond		Covalent bond *within* the molecule and IMF *between* molecules					
Nature of chloride	Neutral	Weakly acidic	Acidic					
Chemical formula of chloride	NaCl	$MgCl_2$	$AlCl_3$	$SiCl_4$	PCl_5	SCl_6	Cl_2	

 Q Both NaCl and $MgCl_2$ contain the electron-rich Cl^- ion, so why is the Cl^- ion not basic in nature?

A: No doubt the Cl^- ion is relatively electron rich, but unfortunately, it is not so electron rich that it can act as a base!

 Q Then why are $MgCl_2$ and $AlCl_3$ acidic in nature?

A: The acidic natures of these two compounds arise from the high charge densities of the Mg^{2+} and Al^{3+} ions. Due to their high charge density, when these ions attract the water molecules, the attractive force is so strong that the *intramolecular* covalent bonds within the H–O–H molecules are so weakened that they break up easily to release H^+ ions.

 Q So, this does not occur for Na^+ of NaCl because its charge density is not high enough to do it?

A: Brilliant! That is why Na^+ in water is neutral in nature.

> **Q** Since phosphorous is in the same group as nitrogen, can phosphorous form PCl_3?

A: Yes, certainly. PCl_3 does exist. If you react phosphorous with just enough chlorine to form PCl_3, then you would get PCl_3. Excess chlorine would convert the PCl_3 to PCl_5.

> **Q** Can PCl_6 be formed?

A: Nope! The maximum oxidation state that the element can have, corresponds to the number of valence electrons it has, which is the (Group number -10) for elements that come from Groups 13 to 18!

> 3. (a) An element **Y** forms the Y^{2-} ion.
>
>   (i) To which group of the Periodic Table does the element **Y** belong?

Explanation:

Using the octet rule as a guideline, since **Y** has taken in two extra electrons, **Y** originally should have six electrons in its valence shell. Therefore, **Y** is in Group 16.

> **Q** Why did you say "using the octet rule as a guideline?"

A: Well, the octet rule is not a "dictating" rule! Not every element would have eight electrons in its valence shell when the element forms a compound.

> **Q** Then, what really explains why a particular compound is formed and not otherwise?

A: A compound is formed because it is the most stable pathway for it to go. Such a compound has the lowest energy possible. So, it is the lower energy level of this particular compound that explains its formation.

(ii) The Y^{2-} ion contains 18 electrons. Write the electronic configuration of Y^{2-}.

Explanation:

The electronic configuration of Y^{2-} is 2.8.8.

(iii) To which period of the Periodic Table does element **Y** belong?

Explanation:

There are three electronic shells in **Y**, so **Y** is from Period 3.

(iv) If element **W** belongs to the same group as element **Y** but the atom of **W** is smaller in size than the atom of **Y**, predict the formula of the ion which **W** forms and give the electronic configuration of **W**.

Explanation:

Since **W** is in the same group as **Y**, **W** would also have six valence electrons. Thus, the formula of the ion which **W** forms is W^{2-}.

As **W** is smaller in size than the atom of **Y**, **W** is above **Y** in the same group. Thus, **W** would have only two electronic shells and the electronic configuration of **W** is 2.6.

Q   Why can't **W** have only one electronic shell?

A: If **W** has only one electronic shell, then this electronic shell is also the valence shell. There should then be six electrons in the valence shell, but the maximum number of electrons in the first electronic shell is only two.

(v) Which is more electronegative, **Y** or **W**?

Explanation:

Since **W** is above **Y** in the group, the valence electrons of **W** are nearer to the nucleus as compared to **Y**. Thus, the valence electrons of **W** would be more strongly attracted to the nucleus than those in **Y**. Hence, **W** is more electronegative than **Y**.

(vi) Which element requires a lower amount of energy to remove its valence electron, **Y** or **W**?

Explanation:

Since **W** is above **Y** in the group, the valence electrons of **W** are nearer to the nucleus as compared to **Y**. Thus, the valence electrons of **W** would be more strongly attracted to the nucleus than those in **Y**. Hence, **Y** requires a lower amount of energy to remove its valence electrons.

Q   So, which element has a smaller atomic radius?

A: By nature, since **Y** has one more electronic shell than **W** and this extra electronic shell is farther away from the nucleus, the atomic radius of **Y** is larger.

Q   So, the atomic size increases down a group because as we go down a group, the valence shell is by default farther away from the nucleus?

A: Brilliant, that is right!

(vii) Explain which element, **Y** or **W**, is a stronger oxidizing agent.

Explanation:

An oxidizing agent itself undergoes reduction by taking in electrons into its valence shell. Thus, the stronger oxidizing agent would be the one that can attract more strongly on its valence electrons. Since the valence shell of **W** is closer to the nucleus than **Y**, **W** is a stronger oxidizing agent.

(b) With reference to the Periodic Table, answer the following questions:

    (i) Suggest the formula of the compound formed between gallium and fluorine. Draw the dot-and-cross diagram of this compound.

Explanation:

Gallium is from Group 13 and fluorine is from Group 17. Using the octet rule as a guideline, Ga would form Ga^{3+} while F would form F^-. The formula is GaF_3.

$$\left[Ga \right]^{3+} 3\left[:\overset{\bullet\times}{\underset{\bullet\bullet}{F}}: \right]^{-}$$

(ii) Draw a dot-and-cross diagram to show the outer shell electrons in a molecule of oxygen difluoride, OF_2.

Explanation:

$$:\overset{\bullet\bullet}{\underset{\bullet\bullet}{F}}\times\overset{\bullet\bullet}{\underset{\bullet\bullet}{O}}\times\overset{\bullet\bullet}{\underset{\bullet\bullet}{F}}:$$

(iii) Explain why the relative atomic mass of chlorine is not a whole number.

Explanation:

The relative atomic mass is the average mass of one atom of the element relative to 1/12 of the mass of a carbon-12 atom. The relative atomic mass is a weighted average of the masses of the different types of isotopes that may be present for a particular element. Thus, the relative atomic mass of chlorine is not a whole number because the ^{35}Cl and ^{37}Cl isotopes are not in a molar composition of 1:1.

Do you know?

— Chlorine-35 and chlorine-37 have relative isotopic masses of 35 and 37, respectively. If a sample of chlorine consists of only chlorine-35 atoms then its relative atomic mass (A_r) would be 35, and if it consists of only chlorine-37 atoms, its relative atomic mass would be 37. A 50:50 mixture of chlorine-35 and chlorine-37 would give chlorine a relative atomic mass of 36. In nature, the isotopes occur in the ratio of about 3:1, i.e., 75% of chlorine-35 and 25% chlorine-37. This results in a relative atomic mass for chlorine of 35.5, as provided in the Periodic Table. Thus, the relative atomic mass is a *weighted average* of the relative isotopic masses of the different isotopes:

$$A_r \text{ of chlorine} = \frac{75(35) + 25(37)}{(75 + 25)} = 35.5.$$

Did you notice that the relative atomic mass is closer to the relative isotopic mass of the isotope that has a *higher proportion* in the mixture of isotopes?

(c) Sulfur dichloride (SCl_2), bromomethane (CH_3Br), and krypton (Kr) are all gases at room temperature.

   (i) Draw the dot-and-cross diagrams of sulfur dichloride (SCl_2) and bromomethane (CH_3Br).

Explanation:

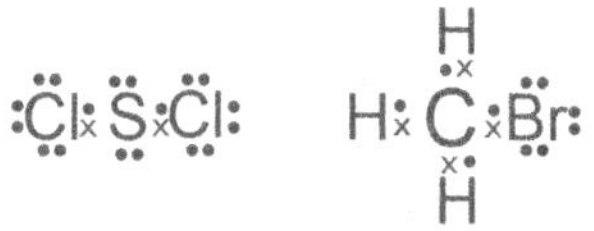

(ii) Which of the three gases will diffuse at the slowest rate at room temperature? Explain your answer based on the Kinetic Theory.

Explanation:

Based on the Kinetic Theory, at the same temperature, each of the three different gases would have the same average kinetic energy per particle. The relative molecular masses of sulfur dichloride (SCl_2), bromomethane (CH_3Br), and krypton (Kr) are respectively, 103.1, 94.9, and 83.8. As kinetic energy is, K.E. $= \frac{1}{2}mv^2$, with a greater mass, the speed would be slower. Hence, SCl_2 will diffuse at the slowest rate at room temperature since it is the heaviest particle.

Do you know?

— According to the Kinetic Theory:
- Matter consists of a large number of small particles, which can be atoms, ions, or molecules.
- There is a large separation between these particles, be it in the solid, liquid, or gaseous state. As such, the size of the particle is *negligible* as compared to the distance of separation between the particles.
- The particles are in constant motion.
- As a consequence of motion, the particles possess *kinetic energy*, which is the energy of motion. The faster the speed of the particle, the higher the kinetic energy.

(Continued)

(Continued)

- The kinetic energy is transferred between particles during their collisions or onto the wall of the container.
- The average kinetic energy of the particles in the system is directly proportional to the temperature of the system.
- The collision on the wall of container gives rise to the concept of pressure.

(d) Sulfur dioxide can be prepared in the laboratory by adding dilute hydrochloric acid to sodium sulfate(IV) (Na_2SO_3).

(i) Determine the oxidation state of the sulfur atom in Na_2SO_3.

Explanation:

Let the oxidation state of sulfur atom be x.
Therefore, $2(+1) + x + 3(-2) = 0$ {As Na_2SO_3 is electrically neutral.}
$$\Rightarrow x = +4.$$

Do you know?

— In order to accurately obtain the oxidation number of an atom, it would be best to know the molecular structure of the species. For example, the oxidation number that is calculated for the S atom in $S_2O_3^{2-}$ is on the average, $2x + 3(-2) = -2 \Rightarrow x = +2$. But the actual oxidation state for each of the S atom is 0 and +4, respectively:

(ii) Give an ionic equation for this reaction.

Explanation:

$$SO_3{}^{2-}(aq) + 2H^+(aq) \rightarrow SO_2(g) + H_2O(l).$$

Q Isn't the above reaction very similar to the reaction of a carbonate with acid?

A: Yes! The reaction of a carbonate ion with an acid is as follows:

$$CO_3{}^{2-}(aq) + 2H^+(aq) \rightarrow CO_2(g) + H_2O(l).$$

(iii) Suggest a chemical test for sulfur dioxide gas. State the observation.

Explanation:

Sulfur dioxide is a reducing agent that can be tested using acidified $K_2Cr_2O_7(aq)$. If the orange $Cr_2O_7{}^{2-}$ solution turns green, then $Cr_2O_7{}^{2-}$ has been reduced to Cr^{3+}. This means that the unknown substance has caused the reduction and hence contains a reducing agent.

Q Can we use acidified $KMnO_4$ to test for SO_2?

A: Yes! If the purple $MnO_4{}^-$ solution turns colorless, then $MnO_4{}^-$ has been reduced to Mn^{2+}. This means that the unknown substance has *caused* the reduction and hence contains a reducing agent.

(iv) The chemical test is based on a specific chemical property of sulfur dioxide. What is this property?

Explanation:

SO_2 is a reducing agent.

(v) Explain why sulfur dioxide is an environmental pollutant.

Explanation:

The accumulation of sulfur dioxide in the air is a source of acid rain which damage buildings, aquatic life, and plants. It is also a potent irritant to the eyes and lungs.

(vi) State one industrial use of sulfur dioxide.

Explanation:

Sulfur dioxide can be further oxidized to SO_3 and then be converted to sulfuric acid, H_2SO_4.

(e) Phosphorous reacts with oxygen to produce an oxide of the formula, P_4O_6.
   (i) Determine the oxidation state of the phosphorous atom in P_4O_6.

Explanation:

Let the oxidation state of phosphorous atom be x.
Therefore, $4(x) + 6(-2) = 0$ {As P_4O_6 is electrically neutral.}
   $\Rightarrow x = +3$.

(ii) Construct a chemical equation for the formation of P_4O_6.

Explanation:

$$P_4(s) + 3O_2(g) \rightarrow P_4O_6(s).$$

(iii) Explain why P_4O_6 has a low melting point.

Explanation:

P_4O_6 is a simple molecular compound with weak intermolecular forces between the P_4O_6 molecules. Hence, P_4O_6 has a low melting point.

(iv) Explain why P_4O_6 does not conduct electricity even in the molten state.

Explanation:

In the molten state, the P_4O_6 exists as discrete molecules. There are no charge carriers. Hence, the liquid is electrically non-conducting.

(v) Explain why P_4O_6 is acidic in nature.

Explanation:

As the phosphorous atom is bonded to highly electronegative oxygen atoms, the phosphorous atom in P_4O_6 is electron deficient. Hence, the phosphorous atom in P_4O_6 can be attacked by water molecules to give the phosphoric(III) acid, H_3PO_3, which is acidic in nature.

> (vi) P_4O_6 can be converted to P_4O_{10} with excess oxygen. Determine the oxidation state of phosphorous in P_4O_{10} and explain whether phosphorous can have another oxidation state that is greater than this.

Explanation:

Let the oxidation state of phosphorous atom be x.
Therefore, $4(x) + 10(-2) = 0$ {As P_4O_{10} is electrically neutral.}
$$\Rightarrow x = +5.$$
Phosphorous cannot have another oxidation state that is greater than this because phosphorous atom has only five valence electrons. The maximum oxidation state that is possible for an element corresponds to the number of valence electrons it has.

 What do you get when P_4O_{10} is dissolved in water?

A: Well, you get phosphoric(V) acid, H_3PO_4.

> 4. Chlorine is an element in Group 17 of the Periodic Table. Chlorine reacts with aqueous potassium iodide to form potassium chloride and iodine.
>   (a) Give the observations when chlorine reacts with potassium iodide.

Explanation:

The solution would turn yellow and black iodine crystals would be observed.

Q Why are there iodine crystals?

A: Iodine is not very soluble in water. Thus, when too much iodine is formed, the solid crystal would precipitate out.

Q Why is iodine not very soluble in water?

A: Iodine is a non-polar molecule with instantaneous dipole–induced dipole (id–id) interaction between the molecules. Water molecules have strong hydrogen bonding between the molecules. If iodine interacts with the water molecules, the amount of energy evolved through these interactions cannot compensate for the energy that is needed to overcome the strong hydrogen bonding between the water molecules.

Q So, is it right to say that ethanol mixes well with water because the hydrogen bonds that are formed between the ethanol and water molecules can release a sufficient amount of energy to compensate for the energy that is needed to overcome the hydrogen bonding between the water molecules?

A: Yes, you are right! This is the explanation behind the maxim "*like dissolves like.*"

Do you know?

— Group 17 elements are strong oxidizing agents. As a result of their high electronegativity, they tend to accept electrons readily from other substances and become reduced. These elements would then form the halide ions, which carry a charge of –1 and has a stable octet configuration in the valence shell.

$$X_2 + 2e^- \rightarrow 2X^-, \quad X = F, Cl, Br, or I.$$

(*Continued*)

(Continued)

— Since electronegativity *decreases* down the group, it is expected that the *ease* in accepting electrons also decreases. In other words, the oxidizing power of the Group 17 elements decreases down the group, with fluorine being the strongest and iodine the weakest oxidizing agent in the group.

— One important reaction to show the decrease in oxidizing power down Group 17 is through the *displacement* reaction. In a displacement reaction, the halogen which is a stronger oxidizing agent, will be able to oxidize the halide ion of the weaker oxidizing agent that is below itself in the group. The halide ion is said to be displaced from the solution. We can make use of this type of reaction to identify the presence of Br^- and I^- ions in solution. Simply add aqueous Cl_2 followed by an organic solvent. Upon shaking the mixture, the presence of Br^- or I^- ions can be confirmed as Br_2 and I_2 provides distinct coloration in the organic layer.

$$Cl_2(aq) + 2Br^-(aq) \rightarrow 2Cl^-(aq) + Br_2(aq);$$

and

$$Cl_2(aq) + 2I^-(aq) \rightarrow 2Cl^-(aq) + I_2(aq).$$

If we try to react I_2 with Cl^-, no reaction will occur. I^- and Cl_2 are not produced!

— Similarly, if a halogen is less likely to undergo reduction, the corresponding halide would be more likely to undergo oxidation. This is easy to understand as reduction means taking in of electrons. If an atom is less likely to take in electrons, it can only mean that the extra electrons are less strongly attracted, which would also indicate that it is more likely that an electron is removed from this same atom. Thus, the weaker the oxidizing agent, the less likely it will be reduced. Conversely, the reduced form of this oxidizing agent will be a stronger reducing agent!

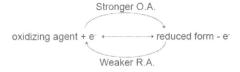

(b) Suggest a test for the presence of iodine.

Explanation:

The presence of iodine can be verified using starch solution. If it turns blue-black, this means that iodine is present.

(c) State the roles of chlorine and iodide in the above reaction.

Explanation:

Chlorine is being reduced; hence, it is acting as an oxidizing agent. Iodide is being oxidized; hence it is acting as a reducing agent.

(d) Give two equations that reflect the roles of chlorine and iodide in part (c).

Explanation:

Oxidation half-equation: $2I^- \rightarrow I_2 + 2e^-$.
Reduction half-equation: $Cl_2 + 2e^- \rightarrow 2Cl^-$.

(e) Construct an ionic equation for the reaction between chlorine and potassium iodide.

Explanation:

$$Cl_2(aq) + 2I^-(aq) \rightarrow 2Cl^-(aq) + I_2(aq).$$

(f) If iodine is added to aqueous potassium chloride, would you expect a similar reaction to occur? Explain.

Explanation:

Chlorine is a stronger oxidizing agent than iodine, i.e., the Cl atom prefers to form Cl^- more than the I atom preferring to form I^-. Thus, Cl^- will not be oxidized by I_2.

The Cl atom prefers to form Cl^- more than the I atom preferring to form I^-. Is this because the extra electron is more strongly attracted in Cl^- than in I^- which is due to the fact that the valence shell of Cl^- is nearer to the nucleus than I^-?

A: Excellent! So, it also means that it is easier to remove an electron from I or I^- than from Cl or Cl^-. This is because the valence shell of I or I^- is farther away from the nucleus.

(g) Astatine is another element in Group 17, but its properties are not well studied because of its radioactive nature.
  (i) Predict with reasons, whether astatine will react with aqueous potassium bromide.

Explanation:

Since astatine is below bromine, astatine is a weaker oxidizing agent than bromine. Astatine will not react with aqueous potassium bromide.

(ii) How would you expect the rate of reaction of astatine with hydrogen to be like as compared to those of iodine and hydrogen?

Explanation:

The oxidizing power of the Group 17 elements decreases down the group. As astatine is lower in the group, the rate of reaction of astatine with hydrogen is slower as compared to those between iodine and hydrogen.

(iii) Construct an equation for the reaction of astatine with potassium.

Explanation:

$$At(s) + K(s) \rightarrow KAt(s).$$

(iv) If solid iodine is black in color, suggest the color for solid astatine.

Explanation:

If solid iodine is black in color, astatine would also be black in color.

Q Wow! The previous question actually shows us that we can make use of group trend to predict the physical and chemical properties of other members in the same group that have not been well studied?

A: Yes, this is the power of the Periodic Table. Elements with similar physical and chemical properties are arranged within the same group. Even across a period, there are also trends of changing physical and chemical properties, which are predictable.

Do you know?

— Group 17 elements exist as simple discrete diatomic molecules that are non-polar with only weak id–id interactions between the molecules. The presence of such weak intermolecular forces of attraction accounts for the relatively low values of the elements' physical properties such as melting point, boiling point, and density.

— Moving down the group from fluorine to iodine, there is a gradual increase in the magnitude of these properties. For example, in observing that fluorine and chlorine are gases, bromine is a liquid, and iodine is a solid at room temperature conditions, we can easily deduce that down the group,

- both melting point and boiling point increase;
- ΔH_{vap} becomes more endothermic; and
- volatility decreases.

Group 17 element	Physical state at 20°C
Fluorine	pale-yellow gas
Chlorine	yellowish-green gas
Bromine	reddish-brown liquid
Iodine	dark violet solid (looks black)

Do you notice that the melting points of Group 17 elements increase down the group BUT those of Group 1 elements actually decrease?

Q   What is volatility?

A: It refers to the changing of phase from a solid or liquid to a gas. Volatility and boiling point have a reciprocal relationship. When we say that ethanol is highly volatile, it also means that it has a low boiling point. So, volatility is related to the strength of the electrostatic attractive force between the particles. A higher volatility means a weaker attractive force.

Q What is the enthalpy change of vaporization (ΔH_{vap})?

A: The enthalpy change of vaporization (ΔH_{vap}) is a measure of volatility. The more endothermic the ΔH_{vap}, the greater the amount of energy needed for a substance to be vaporized (converted to the gaseous state), and thus the substance is said to have low volatility.

Q Since the Group 17 elements are non-polar, does that mean that they are highly soluble in a non-polar solvent?

A: Due to their non-polar nature, the halogens are more soluble in non-polar organic solvents, such as hexane. But they are less soluble in polar solvents of which water is an example. Particularly, when bromine and iodine are each dissolved in an organic solvent, they will produce distinct colorations as shown:

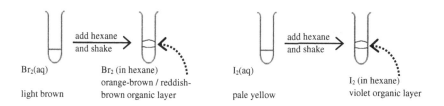

$Br_2(aq)$	Br_2 (in hexane)
	orange-brown / reddish-
light brown	brown organic layer

$I_2(aq)$	I_2 (in hexane)
pale yellow	violet organic layer

add hexane and shake

add hexane and shake

(f) Magnesium reacts with chlorine to form magnesium chloride. Draw the dot-and-cross diagram to show the electronic structures and charges of both ions present in magnesium chloride.

Explanation:

$$\left[Mg \right]^{2+} 2 \left[:\ddot{\underset{..}{Cl}}: \right]^{-}$$

(g) Silver chloride is an insoluble salt. Outline the preparation of pure, dry silver chloride, starting from solid silver nitrate.

Explanation:

Step 1: Prepare an aqueous solution of $AgNO_3$ by dissolving solid $AgNO_3$ in water.

Step 2: Mix the two solutions, $NaCl(aq)$ and $AgNO_3(aq)$, together.

$$AgNO_3(aq) + NaCl(aq) \rightarrow AgCl(s) + NaNO_3(aq).$$

Step 3: Filter the mixture and collect the residue.

Step 4: Wash the residue with deionized water.

Step 5: Dry it and you get the salt crystals!

(h) Chlorine is also used to form chlorofluorocarbon (CFC) products such as $C_2F_3Cl_3$, which contains a C–C single bond.

(i) Give the dot-and-cross diagram of $C_2F_3Cl_3$.

Explanation:

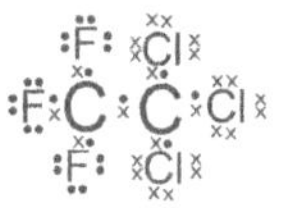

(ii) State one environmental problem associated with the molecule $C_2F_3Cl_3$.

Explanation:

It has been found that CFC products, which are widely used as aerosol propellant or coolants in refrigerators, catalyze the breaking down of ozone. This means that the formation of ozone is actually slower than the rate of it being broken down. When the CFC molecules reach the stratosphere, it breaks up to give *chlorine atoms* which accelerate the decomposition of ozone. As a result, there is an ozone "hole" in the atmosphere, which allows too much ultraviolet light to reach the Earth's surface.

Q Why don't CFC compounds break down before they reach the strato-sphere, but instead only do so when they are in the stratosphere?

A: This is because the C–Cl and C–F bonds are strong enough to allow the CFC compounds to "survive" their journey through the atmosphere to the strato-sphere. In the stratosphere, the molecules break down because of the highly energetic ultraviolet radiation which provides the molecules with a sufficient amount of energy for bond cleavage to take place.

5. This question is about the chemistry of the elements in Period 3 of the Periodic Table.

   (a) Compare the reactions of sodium and of magnesium with cold water. In each case, identify the products formed.

Explanation:

Sodium reacts violently with cold water to produce NaOH(aq) and H_2(g). The reaction is so exothermic that the H_2 gas would be ignited.

$$2Na(s) + 2H_2O(l) \rightarrow 2NaOH(aq) + H_2(g).$$

Magnesium reacts very slowly with cold water because the layer of the insoluble $Mg(OH)_2$ covers the surface and prevents further reaction.

$$Mg(s) + 2H_2O(l) \rightarrow Mg(OH)_2(s) + H_2(g).$$

Do you know?

— Water is a weak acid. It does not react with all types of metal. The tem-perature of the water during the acid–metal reaction would also deter-mine what types of metal are able to react with the water. Hence, when we look at a water–metal reaction, an important factor to consider is the use of cold water or steam.

(Continued)

(Continued)

Cold water:

We know that temperature speeds up the rate of a chemical reaction by providing particles with more kinetic energy to surmount the activation energy. So, we would expect cold water to be less reactive than water at room temperature. Thus, because of this, we would expect that only the more reactive metals, such as potassium, sodium, calcium, and magnesium, are able to react with cold water to give the metal hydroxide and hydrogen gas:

$2K(s) + 2H_2O(l) \rightarrow 2KOH(aq) + H_2(g)$	*Very violent* reaction such that H_2 gas would be ignited.
$2Na(s) + 2H_2O(l) \rightarrow 2NaOH(aq) + H_2(g)$	*Violet reaction* such that H_2 gas would be ignited.
$Ca(s) + 2H_2O(l) \rightarrow Ca(OH)_2(aq) + H_2(g)$	Reacts *readily*.
$Mg(s) + 2H_2O(l) \rightarrow Mg(OH)_2(s) + H_2(g)$	Reacts *very slowly* because the layer of the insoluble $Mg(OH)_2$ covers the surface and prevents further reaction.

The rest of the metals, Al, Zn, Fe, Pb, Cu, Ag, and Au, do not react with cold water!

Steam:

Steam is made up of water molecules ($H_2O(g)$) containing more kinetic energy; hence, it is more reactive than cold water! Other than the more reactive metals, steam can also react with the following metals:

$Mg(s) + H_2O(g) \rightarrow MgO(s) + H_2(g)$	Hot Mg reacts with steam *violently* to give a white glow of light.
$2Al(s) + 3H_2O(g) \rightarrow Al_2O_3(s) + 3H_2(g)$	Reacts very *slowly* because the layer of the insoluble Al_2O_3 covers the surface and prevents further reaction.
$Zn(s) + H_2O(g) \rightarrow ZnO(s) + H_2(g)$	Hot Zn reacts with steam *readily*.
$3Fe(s) + 4H_2O(g) \rightarrow Fe_3O_4(s) + 4H_2(g)$	Red-hot Fe reacts with steam *slowly*.

Q When the metal reacts with steam, why is the metal oxide formed instead of the metal hydroxide?

A: When the metal reacts with steam, the *temperature* is high. Hence, the metal hydroxide undergoes dehydration to form the metal oxide:

$$M(OH)_2(s) \rightarrow MO(s) + H_2O(l).$$

Q If Al can react with steam, how can we use Al metal for cookware?

A: The highly insoluble Al_2O_3 that is formed would prevent the Al metal beneath from further reaction. This is known as chemical passivation, as the metal becomes "passive" or unreactive.

(b) Draw the electronic structures, including the charges, of the ions present in magnesium oxide.

Explanation:

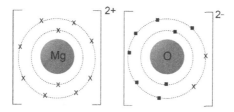

Q What is the difference between an electronic structure and dot-and-cross diagram?

A: An electronic structure shows all the electronic shells, including the inner-core electrons, of the species. A dot-and-cross diagram only shows the valence electrons.

(c) Magnesium oxide has a very high melting point. Explain.

Explanation:

Magnesium oxide is an ionic compound with strong ionic bonds between the oppositely charged ions. Hence, more energy is required to overcome the ionic bonds, accounting for its high melting point.

(d) State one important use of magnesium oxide because of the property mentioned in part (c).

Explanation:

Due to the high melting point of magnesium oxide, it is commonly used as furnace lining due to its high refractory property.

(e) Compare the reactions of magnesium and of aluminum with chlorine. In each case, identify the products formed.

Explanation:

Magnesium reacts vigorously with chlorine to give the ionic solid, $MgCl_2$:

$$Mg(s) + Cl_2(g) \rightarrow MgCl_2(s).$$

Aluminum reacts vigorously with chlorine to give $AlCl_3$, which is ionic in solid state but transforms to the simple molecular form when melted:

$$Al(s) + 3/2Cl_2(g) \rightarrow AlCl_3(s).$$

(f) Draw the dot-and-cross diagram of aluminum chloride, assuming it is simple molecular form.

Explanation:

$$\overset{\overset{\displaystyle :\ddot{Cl}:}{\underset{\displaystyle :\ddot{Cl}:}{}}{\underset{x}{\overset{x}{}}} Al \times \ddot{Cl}:$$

> **Q** Why is $AlCl_3$ able to undergo dimerization in the molten state as shown below?
>
>
>
> Dative covalent bond

A: The Al atom of $AlCl_3$ does not have an octet configuration. Hence, through dative covalent bonding, each Al atom of the Al_2Cl_6 dimer would then adopt the octet configuration.

> (g) It is said that "a metal reacts with a non-metal to give ionic compound." Explain, in terms of structure and bonding, why the statement is not necessary true for aluminum chloride.

Explanation:

Although aluminum is a metal while chlorine is a non-metal, aluminum chloride is a covalent compound with covalent bonds between the Al and Cl atoms when in the molten state.

> **Q** Why is $AlCl_3$ a simple molecular compound whereas AlF_3 is an ionic compound in the molten state?

A: Whether the aluminum compound is ionic or not depends very much on the counter anion. What do we mean by that? If an Al^{3+} ion is near a Cl^- ion, due to the high charge density ($\propto \frac{q_+}{r_+}$) of the Al^{3+} ion and the high polarizability (ability to be polarized) of the Cl^- ion, the electron cloud of the Cl^- ion would be distorted to an extent that when you observe the molten $AlCl_3$, you would find that it is a simple molecular compound, and not an ionic one! But now, if the Al^{3+} is near a F^- ion, due to the low polarizability of the F^- ion, there would be very minimal distortion of electron density of the F^- ion. Hence, you would observe AlF_3 as an ionic compound.

(h) Pure sand is silicon(IV) oxide. It has a macromolecular structure similar to that of diamond. Suggest two physical properties of silicon(IV) oxide.

Explanation:

Due to the strong Si–O covalent bonds in SiO_2, we would expect SiO_2 to be hard and have a high melting point as it is difficult to break the strong covalent bonds.

In addition, as it does not contain mobile electrons, so we would expect SiO_2 to be electrically non-conducting.

(i) Chlorine(VII) oxide, Cl_2O_7, has a simple molecular structure.
  (i) Suggest one physical and one chemical property of Cl_2O_7.

Explanation:

Since Cl_2O_7 has a simple molecular structure, it would have a low melting point due to the weak intermolecular forces between the Cl_2O_7 molecules.

As Cl_2O_7 is a non-metal oxide, it should be acidic in nature. Hence, it can react with a base.

Q Would you expect Cl_2O_7 to be an oxidizing or reducing agent?

A: The oxidation state of the Cl atom in Cl_2O_7 is +7, which corresponds to the number of valence electrons the Cl atom has. Thus, the Cl atom in Cl_2O_7 cannot be further oxidized. Therefore, Cl_2O_7 is likely to be an oxidizing agent.

(ii) State the oxidation state of the chlorine atom in Cl_2O_7.

Explanation:

Let the oxidation state of chlorine atom be x.
Therefore, $2(x) + 7(-2) = 0$ {As Cl_2O_7 is electrically neutral.}
$$\Rightarrow x = +7.$$

(iii) Explain why the oxidation state that you mentioned in part (i)(ii) is as such.

Explanation:

Cl atom belongs to Group 17; it has seven valence electrons. The maximum oxidation state of the Cl atom in Cl_2O_7 corresponds to the number of valence electrons it has.

CHAPTER 13

METALS AND EXTRACTION

Do you know?

— Most metals are found in the ground as compounds called minerals. Impure minerals are called ores. An *ore* is a naturally occurring source of a metal, from which the metal is extracted via economical means. The ore of a metal is usually in the form of the oxides, sulfides, chlorides, or carbonates.

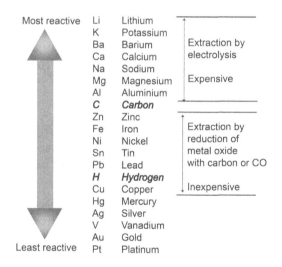

In Chapter 10, we learned that for metals that are above carbon in the reactivity series (K, Na, Ca, Mg, and Al), their oxides cannot be reduced

(Continued)

(Continued)

by carbon back to their metallic form. Therefore, to extract the metal, electrolysis has to be employed. This thus makes the metals expensive. Oxides of Zn, Fe, Pb, and Cu can be reduced by carbon through heating as these metals are below carbon in the reactivity series. The prices of the metals that are extracted through such a method are relatively inexpensive.

(1) Iron is extracted form hematite, which contains iron(III) oxide, using the blast furnace.

    (a) Explain why carbon can be used to extract iron from its oxide ores.

Explanation:

Iron is below carbon in the reactivity series. This makes carbon more likely to undergo oxidation as compared to iron. Hence, carbon can be used to reduce iron(III) oxide into the iron metal.

 Q If the metal is below carbon in the reactivity series, it means that carbon is more likely to undergo oxidation than the metal. So how does this actually enable carbon to reduce the metal oxide?

A: If you put the metal oxide and carbon together, since carbon is above the metal in the reactivity series, then CO_2 is preferred to the metal oxide. The carbon will "extract" the oxygen from the oxide to form CO_2, leaving the reduced metal oxide, which is the metal, behind. Get it? May be the following would be helpful:

$$\text{Metal oxide} + \text{Carbon} \rightarrow \text{Metal} + \text{Carbon dioxide}.$$

Carbon dioxide is preferred over the metal oxide.
The metal is preferred over carbon.
Therefore, the right-hand side of the equation is preferred!

Do you know?

— Iron is extracted from its ore, hematite, using the blast furnace. A mixture of the iron ore, limestone ($CaCO_3$), and coke (carbon) is fed into the blast furnace from the top:

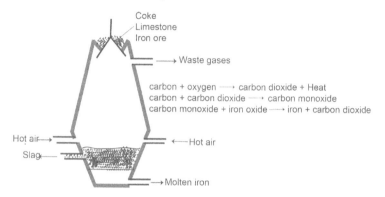

Hot air that has been preheated by the hot waste gases is introduced at the bottom of the furnace. In the blast furnace, there are a few important reactions taking place:

1. *Formation of carbon dioxide*
The burning of coke is highly exothermic, which provides the energy that is needed for the reduction:

$$C(s) + O_2(g) \rightarrow CO_2(g).$$

The decomposition of limestone, $CaCO_3$, also produces CO_2:

$$CaCO_3(s) \rightarrow CaO(s) + CO_2(g).$$

The purpose of the calcium(II) oxide is to help to remove impurities such as sand and clay in the ore, through the following reaction:

$$CaO(s) + SiO_2(s) \rightarrow CaSiO_3(l).$$

The molten slag, $CaSiO_3(l)$, which is lighter than molten iron, can be drained off easily from the bottom of the furnace.

2. *Formation of carbon monoxide*
The carbon dioxide formed from the burning of coke would be reduced to carbon monoxide when it encounters more coke:

$$CO_2(g) + C(s) \rightarrow 2CO(g).$$

(Continued)

(Continued)

3. *Reduction of iron(III) oxide in hematite*

It is the carbon monoxide that is responsible for reducing the iron(III) oxide to iron:

$$Fe_2O_3(s) + 3CO(g) \rightarrow 2Fe(l) + 3CO_2(g).$$

Molten iron is known as pig iron or cast iron.

The waste gases, which include carbon dioxide, carbon monoxide, and others, are released into the atmosphere. This is environmentally unfriendly!

(b) Name and explain the type of chemical reaction involved when iron ore is converted to iron.

Explanation:

When Fe_2O_3 is converted to iron, the oxidation state of Fe^{3+} decreases from +3 to 0. This is a reduction process. Thus, there must be a corresponding oxidation reaction happening at the same time, which is the oxidation of the CO to CO_2. Therefore, the overall reaction is redox in nature.

Q   So, the oxidation state of carbon in CO increases from +2 to +4 in CO_2?

A: You are right!

(c) Write a balanced chemical equation for the reaction between iron(III) oxide and carbon monoxide.

Explanation:

$$Fe_2O_3(s) + 3CO(g) \rightarrow 2Fe(l) + 3CO_2(g).$$

(d) Draw the dot-and-cross diagram of iron(III) oxide.

Explanation:

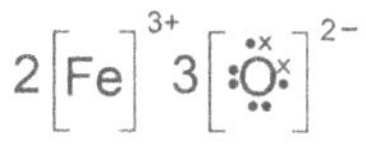

$$2\left[\text{Fe}\right]^{3+} 3\left[\ddot{\overset{\bullet\times}{\text{O}}}\right]^{2-}$$

(e) Draw the dot-and-cross diagram of carbon monoxide if both atoms have achieved the stable octet configuration.

Explanation:

$$:\text{C}\overset{\times\times}{\underset{\times}{:}}\text{O}\times$$

(f) What is the use of limestone in the blast furnace?

Explanation:

The decomposition of limestone, $CaCO_3$, produces CaO and CO_2:

$$CaCO_3(s) \rightarrow CaO(s) + CO_2(g).$$

The purpose of the calcium(II) oxide is to help remove impurities such as sand and clay in the ore, through the following reaction:

$$CaO(s) + SiO_2(s) \rightarrow CaSiO_3(l).$$

The CO_2 is used to oxidize carbon to CO:

$$CO_2(g) + C(s) \rightarrow 2CO(g).$$

(g) The waste gases consist of carbon dioxide, carbon monoxide, and nitrogen. Explain the presence of each of these gases.

Explanation:

Carbon dioxide is produced from the oxidation of carbon by oxygen and also from the decomposition of the $CaCO_3$.

Carbon monoxide is produced from the oxidation of carbon by carbon dioxide.

Nitrogen is from the air that is pumped into the furnace in which the oxygen is used to oxidize the carbon while the nitrogen remains unreacted.

(h) Given that 28 tons (1 ton = 1000 kg) of hematite are loaded into the furnace, calculate

   (i) the minimum mass of coke required for complete extraction of the iron.

Explanation:

Molar mass of $Fe_2O_3 = 2(55.8) + 3(16.0) = 159.6$ g mol^{-1}.

Amount of Fe_2O_3 in 28 tons $= \dfrac{28,000,000}{159.6} = 1.75 \times 10^5$ mol.

$$Fe_2O_3(s) + 3CO(g) \rightarrow 2Fe(l) + 3CO_2(g).$$

One mole of Fe_2O_3 reacts with three moles of CO. Assuming that all the CO that is needed come from coke (C), then 1.75×10^5 mol of Fe_2O_3 needs $= 1.75 \times 10^5 \times 3 = 5.25 \times 10^5$ mol of CO.

Therefore, the minimum mass of coke that is required for the complete extraction of the iron $= 5.25 \times 10^5 \times 12.0 = 6.30 \times 10^6$ g $= 6.30 \times 10^3$ tons.

Q Why do you need to assume that all the CO that is needed comes from coke?

A: In addition to the CO_2 from the burning of the coke, the CO_2 that is produced from the decomposition of $CaCO_3$ can also help to convert the carbon to carbon monoxide as follows:

$$CO_2(g) + C(s) \rightarrow 2CO(g).$$

(ii) the maximum possible mass of iron extracted.

Explanation:

Amount of Fe extracted $= 2 \times$ Amount of Fe_2O_3 in 28 tons $= 3.50 \times 10^5$ mol.

Therefore, the maximum possible mass of iron extracted $= 3.50 \times 10^5 \times 55.8 = 1.95 \times 10^7$ g.

(i) Given that only 18 tons of iron were retrieved from part (h), calculate the percentage purity of the hematite.

Explanation:

Amount of Fe in 18 tons $= \dfrac{18{,}000{,}000}{55.8} = 3.22 \times 10^5$ mol.

Amount of $Fe_2O_3 = \dfrac{1}{2} \times$ Amount of Fe in 28 tons $= 1.61 \times 10^5$ mol.

Mass of $Fe_2O_3 = 1.61 \times 10^5 \times 159.6 = 2.57 \times 10^7$ g $= 25.7$ tons.

Percentage purity of hematite $= \dfrac{25.7}{28.0} \times 100 = 91.8\%$.

(j) Iron is often alloyed to reduce rusting.

   (i) Explain the term *alloy*.

Explanation:

A pure metal by itself is soft, thus they are malleable and ductile. But if we introduce another metal atom of a different size into the *regular* lattice structure of a metal, we can increase the strength of the metal. Such a solid solution of metals is known as an *alloy*. Sometimes, instead of introducing another metal atom, carbon atoms are used instead.

Why is the strength of an alloy much stronger than that of a pure metal?

A: As the size of the atom of the impurity is different from that of the metal, it disrupts the regular arrangement of the atoms in the pure metal. Thus, when a force is applied, the atom of the impurity, which is bigger in size, would prevent the layers of atoms from sliding over one another. This makes the metal harder!

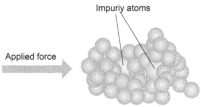

Q How is an alloy formed?

A: An alloy is usually made by melting different metal components together and then solidifying the mixture. As such, the metal components of an alloy must be soluble in one another when in the molten state, and they should not separate into distinct layers when solidified.

Do you know?

— Properties that an alloy has but are absent in a pure metal:

- An alloy is usually *harder*. For instance, brass, which is a composite of zinc and copper, is harder than its constituent metals.
- An alloy can be more *resistant* to corrosion. For instance, stainless steel, which is iron doped with other atoms such as carbon, chromium, manganese, etc., is more resistant to corrosion than pure iron itself.
- An alloy can have a *lower* melting point than the pure metal. For instance, solder, which is an alloy of tin and lead, has a lower melting point than its constituent metals.
- An alloy can have a better appearance than its constituent metals. For instance, pewter is an alloy of tin, copper, and antimony. It is aesthetically more appealing than the dull gray tin.

(ii) Name one element besides carbon that is used for this purpose.

Explanation:

Manganese or chromium.

(iii) Explain why an alloy can be more resistant to corrosion than a pure metal.

Explanation:

The metal that is used to dope the pure metal may be more likely to undergo corrosion than the pure metal itself. This thus protects the pure metal from undergoing corrosion.

Do you know?

— Other than using alloying to minimize corrosion, the common methods for rust prevention are:

Painting
Coating iron with a layer of paint excludes water and oxygen from coming in contact with the iron; this helps to prevent corrosion.

Galvanization
By coating a layer of zinc, which corrodes preferentially to the iron due to its higher reactivity (refer to Chapter 10 on Electric Cells and the Reactivity Series), the iron would be protected.

Cathodic protection
A piece of more reactive metal, such as magnesium, is connected to the iron. As the magnesium preferentially undergoes oxidation, it is acting

(Continued)

(Continued)

as an anode. The electrons from this oxidation process would flow to the iron. This makes the iron electron-rich, in which only reduction can happen there. Hence, the iron is acting as a cathode where it cannot undergo oxidation. As time passes, the magnesium rod disappears and is replaced with a new one. The magnesium metal is known as the *sacrificial* metal. This method is commonly used to protect iron pipes.

(iv) Iron is malleable. Describe how this property can be explained in terms of its structure.

Explanation:

The atoms in iron are held together by metallic bonds which are strong and non-directional. The metallic bonds are the electrostatic attractive forces between the positive ions and the delocalized electrons. Thus, when a force is applied to a piece of iron, the atoms simply move over one another without breaking of the metallic bonds. This thus accounts for the malleability and ductility of metals.

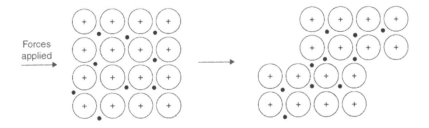

(v) Explain why the strength of an alloy is different from that of a pure metal.

Explanation:

As the size of the atom of the impurity is different from that of the metal, it disrupts the regular arrangement of the atoms of the pure metal. Thus, when a force is applied, the impurity atom which is bigger in size, would prevent the layers of atoms from sliding over one another. This makes the metal harder.

(vi) Explain why iron is a good conductor of heat and electricity.

Explanation:

The solid lattice structure of iron has a sea of delocalized electrons which can act as mobile charge carriers under the influence of an applied electrical potential. This accounts for the good electrical conductivity of the metal.

In addition, because the delocalized electrons are small and light, they can transfer heat energy across a piece of metal more efficiently through collisions among themselves. This accounts for the good thermal conductivity of the metal.

(vii) Explain why hot water tanks used in industries are made of copper and not steel, which is an alloy containing iron.

Explanation:

Copper does not corrode under hot water as easily as steel does. In addition, copper is a better conductor of heat energy than steel, thus a hot water tank that is made of copper can heat up the water more uniformly.

 Q Wait a minute, if we use stainless steel instead, would it be better as corrosion is minimized?

A: Stainless steel does not corrode as easily as steel, but copper is still more resistant to corrosion in the long run.

Q   Why is copper a better conductor of heat than steel?

A: This is because the copper solid lattice has more delocalized electrons than steel. Thus, more heat energy can be transferred faster across a piece of copper of the same dimensions than a piece of steel. This also accounts for the higher electrical conductivity of copper compared to steel.

Q   If copper is better at conducting heat, won't it lose heat more easily than steel?

A: Good thinking! This problem can be solved by using insulation. Thus, overall, it is still better to use copper to construct a hot water tank.

(k) State two other metals that can be extracted using carbon. Hence, explain why aluminum cannot be extracted from aluminum oxide using the method mentioned.

Explanation:

The two metals are tin and zinc.

Since alumimum is above carbon in the reactivity series, this implies that aluminum metal is more likely to undergo oxidation than carbon. This would mean that carbon cannot be used to reduce aluminum oxide.

(l) What technological breakthrough was required before the possible extraction of metallic aluminum?

Explanation:

The melting point of aluminum oxide is simply too high to melt and extract the aluminum economically. With the introduction of cryolite, Na_3AlF_6, the melting point is lowered drastically. This thus allows the aluminum to be extracted from the aluminum oxide through electrolytic means.

Do you know?

— The reactions that are thought to occur at the respective electrodes during the electrolysis of molten Al_2O_3 are as follows:

Anode: $2O^{2-} \rightarrow O_2 + 4e^-$.
Cathode: $Al^{3+} + 3e^- \rightarrow Al$.
Overall ionic equation: $4Al^{3+}(l) + 6O^{2-}(l) \rightarrow 4Al(l) + 3O_2(g)$.

(m) A student suspected that a dilute solution of hydrochloric acid could speed up the rusting of iron. To test his hunch, the student immersed a piece of iron nail in a test tube half-filled with tap water. He then added a few drops of the acid.

(i) Can the student use boiled water in place of the tap water? Explain.

Explanation:

The student cannot use boiled water in place of tap water as boiled water lacks oxygen. This is because the rusting of iron needs both water and oxygen to be present simultaneously.

Do you know?

— The brown rust that is formed is actually a form of hydrated iron(III) oxide:

$$4Fe(s) + 3O_2(g) + 2xH_2O(l) \rightarrow 2Fe_2O_3 \cdot xH_2O(s).$$

> **Q**   Do all metals corrode when expose to air?

A: Both aluminum and chromium are protected from corrosion because of the formation of an impervious layer of oxide. That is why you need to know of the anodization of aluminum and chromium plating. These can help prevent corrosion.

> (ii) What would happen when a layer of oil is spread on the surface of the water in the test tube? Explain.

Explanation:

The layer of oil would exclude oxygen gas from dissolving in the water. Thus, rusting would stop after some time when the oxygen that has already been dissolved in the water is used up.

> (iii) In order to test the reliability of his hunch, a control experiment needed to be set up. Why was this important?

Explanation:

If there is no control experiment, he would not be able to compare the rate of rusting as there would be nothing to compare against.

> (iv) Describe an appropriate set-up for the control experiment.

Explanation:

Since the presence of the acid is the subject of study, an appropriate set-up for the control experiment would be to immerse a piece of iron nail in a test-tube that is half-filled with tap water, but without the addition of the acid. The two set-ups must be placed side-by-side under the same temperature.

(v) Other than excluding the factors that would accelerate rusting, suggest other methods that may be helpful to prevent corrosion.

Explanation:

To prevent corrosion, we can (i) electroplate the metal with chromium; (ii) paint the metal with a layer of paint; (iii) coat a layer of zinc that preferably undergoes corrosion (galvanization); and (iv) connect a piece of more reactive metal, such as magnesium, to the iron (cathodic prevention).

(n) A sample of a compound of iron is analyzed. The sample contains 0.547 g of potassium, 0.195 g of iron, 0.252 g of carbon, and 0.294 g of nitrogen. Calculate the empirical formula of this compound.

Explanation:

Elements present	K	Fe	C	N
Assuming 100 g of mass	54.7	19.5	25.2	29.4
Amount (in mol)	$54.7/39.1 = 1.4$	$19.5/55.8 = 0.35$	$25.2/12 = 2.1$	$29.4/14 = 2.1$
Divide by smallest number of mole	4	1	6	6
Simplest ratio	4	1	6	6

The empirical formula of the unknown compound is $K_4FeC_6N_6$.

2. The metal antimony, Sb, has a proton number of 51. It can be extracted from its oxide via a reaction with carbon in a furnace. The equation for the reaction is:

$$Sb_2O_3(s) + 3C(s) \rightarrow 2Sb(s) + 3CO(g).$$

(a) What is the relationship between the reactivity of the metals and their method of extraction?

Explanation:

Metals that are above carbon in the reactivity series are more likely to undergo oxidation than carbon. Hence, these metals cannot be extracted from their ores by reduction using carbon. So, to extract these metals, electrolysis has to be employed. As for the metals below carbon in the reactivity series, they can be extracted simply through reduction with carbon.

> (b) From the method of extraction of antimony, suggest where the metal might be placed in the reactivity series.

Explanation:

Antimony is below carbon in the reactivity series.

> (c) State the oxidation number for the atoms involved in the above chemical reaction. Hence, state the nature of the above reaction.

Explanation:

$$Sb_2O_3(s) + 3C(s) \rightarrow 2Sb(s) + 3CO(g).$$

Oxidation number: +3 −2   0   0   +2 −2

Sb undergoes reduction as its oxidation state decreases from +3 to 0. Carbon undergoes oxidation as its oxidation state increases from 0 to +2. Hence, the above reaction is a redox reaction.

> (d) What is the maximum amount of antimony, in moles, that can be produced from the reaction of antimony oxide with one mole of carbon?

Explanation:

One mole of carbon would give $\frac{2}{3}$ moles of antimony.

(e) What is the maximum mass of antimony, in tons, that can be formed from 25 tons of its oxide?

Explanation:

Molar mass of $Sb_2O_3 = 2(121.8) + 3(16.0) = 291.6$ g mol^{-1}.

Amount of $Sb_2O_3 = \dfrac{25,000,000}{291.6} = 8.57 \times 10^4$ mol.

Amount of $Sb = 2 \times$ Amount of $Sb_2O_3 = 1.71 \times 10^5$ mol.

Mass of $Sb = 1.71 \times 10^5 \times 121.8 = 2.09 \times 10^7$ g $= 20.9$ tons.

(f) Based on the position of antimony in the Periodic Table, suggest the physical properties of pure antimony.

Explanation:

Antimony is positioned after tin and before iodine in the same period of the Periodic Table. Since tin is a metal while iodine is a non-metal, we would expect antimony to have the characteristics of a metalloid.

(g) State the other possible oxidation state for antimony.

Explanation:

Antimony belongs to Group 15, possessing five valence electrons. Hence, the two possible oxidation states for antimony are +3 and +5. Thus, the other possible oxidation state is +5.

(h) Explain your choice of the oxidation state in part (g).

Explanation:

The maximum oxidation state of an element corresponds to the number of valence electrons it has. This explains the choice of the oxidation state.

> (i) How would you expect the reducing power of antimony to be as compared to phosphorous? Explain.

Explanation:

The reducing power of antimony is greater than that of phosphorous. The valence electrons of the elements that are lower down a group are farther away from the nucleus. Hence, the valence electrons are more easily removed. This would imply that elements that are lower down in a group are easily oxidized, thus having a greater reducing power.

 If the reducing power increases down a group, does that mean that the oxidizing power would decrease down the group?

A: Brilliant! An oxidizing agent undergoes reduction, i.e., takes in electrons. Down a group, since the valence shell is farther away from the nucleus, the extra electrons that an oxidizing agent takes in would be less strongly attracted. Hence, the oxidizing power would decrease down the group!

> (j) How would you expect the oxidizing power of antimony to be as compared to tin? Explain.

Explanation:

An oxidizing agent takes in electrons and thus undergoes reduction. Antimony and tin belong to the same period. Antimony is positioned after tin in the same period. Since the electrostatic attractive force acting on the valence electrons increases across a period, we would expect the extra electrons that antimony takes in to be more strongly attracted than those for tin. Thus, antimony would have a greater oxidizing power than tin.

 If the oxidizing power increases across a period, does that mean the reducing power would decrease across the period?

A: Yes, you are right! A reducing agent undergoes oxidation, i.e., loses electrons. It is thus more difficult to lose electrons across a period.

Do you know?

— To understand whether it is easier to lose electrons across a period of elements, let us look at the electronic configurations of some Period 3 elements:

Element	No. of protons (NC)	Electronic configuration	Net electrostatic attractive force on valence electrons (NC − shielding effect)
Sodium (Na)	11	2.8.1	11 − 10 = +1
Magnesium (Mg)	12	2.8.2	12 − 10 = +2
Aluminum (Al)	13	2.8.3	13 − 10 = +3
Silicon (Si)	14	2.8.4	14 − 10 = +4
Phosphorous (P)	15	2.8.5	15 − 10 = +5
Sulfur (S)	16	2.8.6	16 − 10 = +6
Chlorine (Cl)	17	2.8.7	17 − 10 = +7
Argon (Ar)	18	2.8.8	18 − 10 = +8

Do you see that the *net* electrostatic attractive force acting on the valence electrons has increased across the elements in Period 3? This trend is true even with Periods 2 or 4. So, with an *increase* in the net electrostatic attractive force acting on the valence electrons, we would expect *increasing difficulty to remove valence electrons* across the

(Continued)

(*Continued*)

period. Relatively, it would mean that there is an *increasing affinity for the elements to gain electrons* across the period!

In a nutshell, remember:

- Elements *across a period* possess *increasing* ease in gaining electrons BUT *decreasing* ease in losing electrons.
- Elements *down a group* possess *decreasing* ease in gaining electrons BUT *increasing* ease in losing electron.

3. In the Stone Age, the only metals available were gold and silver. Later, the Bronze Age came about. Bronze is made by mixing copper and tin.

   (a) What is the name given to such a mixture of metals from different elements?

Explanation:

A mixture of metals from different elements is known as an alloy.

(b) Why is bronze used instead of pure copper or tin?

Explanation:

Bronze is probably harder and stronger than pure copper or tin.

Q What is that greenish compound that is coated on a piece of old bronze artifact?

A: It is probably copper(II) carbonate that was formed from the reaction between copper(II) oxide and carbon dioxide in the air:

$$CuO(s) + CO_2(g) \rightarrow CuCO_3(s).$$

(c) Tin was extracted from its ore *cassiterite*, SnO_2, through the process called smelting. In smelting, the ore is heated with carbon.

(i) What type of compound, ionic or covalent, is SnO_2?

Explanation:

SnO_2 is a covalent compound.

Q Why is SnO_2 not an ionic compound?

A: This is another example in which metal and non-metal does not react to give an ionic compound. If SnO_2 is an ionic compound, then it would consist of Sn^{4+} and O^{2-} ions. This would mean that a Sn atom has to lose four electrons. It is simply too energetically demanding to do that. Even if it does, the charge density of Sn^{4+} is too great. Its polarizing power would be very strong. This would then make SnO_2 a covalent compound. Hence, when Sn and O_2 react, SnO_2 has to be a covalent compound.

(ii) Draw the dot-and-cross diagram of SnO_2.

Explanation:

$$\overset{\times\times}{\underset{\times}{\text{O}}}\overset{\times}{\text{:}}\text{Sn:}\overset{\times}{\times}\overset{\times\times}{\underset{\times}{\text{O}}}$$

(iii) Write the chemical equation for the above extraction reaction.

Explanation:

$$C(s) + SnO_2(s) \rightarrow Sn(s) + CO_2(g).$$

(iv) In view of the extraction reaction, where would the position of carbon be in the metal reactivity series relative to tin?

Explanation:

Carbon would be above tin in the reactivity series. This is because metals that are below carbon in the reactivity series are less likely to undergo oxidation than carbon. Hence, these metals can be extracted from their ores by reduction using carbon.

(d) Brass is another alloy consisting of zinc and copper.

(i) Describe, with the aid of a labeled diagram, the structure of a metal such as copper.

Explanation:

The copper metal can be viewed as a rigid lattice of positive ions surrounded by a *sea of delocalized* electrons. What holds the lattice together is the strong metallic bonding — the electrostatic attraction between the positive ions and the delocalized valence electrons.

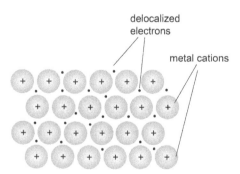

delocalized
electrons

metal cations

(ii) Explain, in terms of their structures, why both zinc and copper are good conductors of electricity.

Explanation:

Both zinc and copper are good conductors of electricity because of the presence of the sea of delocalized electrons which can act as mobile charge carriers when an electrical potential is applied across a piece of the metal.

(iii) A 1.2 g sample of powdered brass was analyzed by reaction with excess dilute sulfuric acid. The zinc reacts as shown in the equation to form 0.072 dm^3 of hydrogen measured at room temperature and pressure:

$$Zn(s) + 2H^+(aq) \rightarrow Zn^{2+}(aq) + H_2(g).$$

(I) Give a name for the above reaction.

Explanation:

The reaction is an acid–metal reaction or a redox reaction.

Q Why does the copper not react with the H^+?

A: Copper is below hydrogen gas in the reactivity series. This means that hydrogen would prefer to be oxidized than copper. Thus, when you mix copper and acid together, the copper cannot reduce the H^+ back to H_2.

Q So, does it mean that we can use hydrogen gas to reduce copper oxide back to copper?

A: Certainly. The concept is similar to the fact that you can use carbon to reduce any metal oxide only if the metals are below carbon in the reactivity series.

> **Q** If copper does not react with acids, why does a piece of copper dumped into nitric acid produce brown fumes?

A: Copper does not react with the H^+ from HNO_3 but it does react with the NO_3^- ion. This accounts for the formation of brown fumes which is actually NO_2 gas. If you dump a piece of copper into HCl or H_2SO_4, there is no acid–metal reaction at all!

> (II) What is the nature of the above reaction?

Explanation:

The reaction is in fact redox in nature. This is because Zn undergoes oxidation as its oxidation state increases from 0 to +2, while H^+ undergoes reduction as its oxidation state decreases from +1 to 0.

> (III) Suggest why brass was used in a powdered form rather than in a lump.

Explanation:

The powdered form ensures a greater rate of reaction between the metal and acid due to the greater surface area of contact between the acid and metal.

> (IV) Calculate the mass of zinc in the sample of brass.

Explanation:

At room temperature and pressure, 1 mole of gas has a volume of 24.0 dm^3.

Amount of H_2 in 0.072 dm^3 = $\dfrac{0.072}{24.0}$ = 3.0×10^{-3} mol.

Amount of Zn = Amount of H_2 in 0.072 dm^3 = 3.0×10^{-3} mol.

Therefore, mass of Zn = $3.0 \times 10^{-3} \times 65.4 = 0.196$ g.

> (V) Calculate the percentage of zinc in the sample of brass.

Explanation:

Percentage of zinc in the sample of brass = $\dfrac{0.196}{1.2} \times 100 = 16.4\%$.

> (VI) Describe how aqueous ammonia can be used to show that only the zinc in the sample reacted with the acid.

Explanation:

If aqueous ammonia is added to the solution slowly till in excess, a white precipitate of $Zn(OH)_2$ would form. In excess aqueous ammonia, the white precipitate would dissolve to give a colorless solution.

Do you know?

— Zinc oxide is an amphoteric oxide. It can react with both an acid and base:
$$ZnO(s) + 2HCl(aq) \rightarrow ZnCl_2(aq) + H_2O(l); \text{ and}$$
$$ZnO(s) + 2NaOH(aq) \rightarrow Na_2ZnO_2(aq) + H_2O(l).$$
$$\text{sodium zincate}$$

> (VII) If you are to study how the rate of the reaction is dependent on the concentration of the acid, with the aid of a labeled diagram, briefly describe how you would do it.

Explanation:

The volume of hydrogen gas can be collected as shown below:

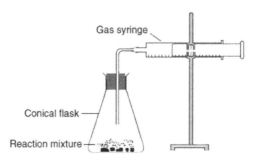

The volume of the hydrogen that is collected can be monitored with respect to time. This would give us the rate of the reaction for different concentrations of the acid used.

4. In the early 19th century, cans made of tin-plated iron were first used to preserve food. Tin has several advantages, including: it corrodes much less readily than iron and it forms a protective oxide layer over the iron. When the tin coating is scratched, however, the iron rusts faster when it is in contact with tin. Fortunately, neither Fe^{2+} nor Sn^{2+} ions are toxic.

   (a) State *two* substances which are necessary for iron to rust and from which iron is protected by the tin layer.

Explanation:

Essentially, iron needs oxygen and water in order to rust. The tin layer would exclude the iron from coming in contact with these two elements for rusting.

   (b) Write a balanced equation for the reaction where the presence of tin ions encourages the iron to corrode.

Explanation:

$$Sn^{2+} + Fe \rightarrow Fe^{2+} + Sn.$$

Q   Why does the Sn^{2+} ion cause the Fe to corrode?

A: From the reactivity series, iron is above tin. This means that iron would prefer to undergo oxidation than tin. Hence, when you have Fe and Sn^{2+} together, the Fe would prefer to become Fe^{2+}, while Sn^{2+} prefers to be formed back to Sn.

(c) Rust is often given the formula $Fe_2O_3 \cdot xH_2O$, where x has a variable non-integral value. Calculate the value of x for a sample of rust which lost 22% mass (as steam) when heated to constant mass.

Explanation:

Assuming 100 g of $Fe_2O_3 \cdot xH_2O$, there are 22 g of H_2O.

Amount of H_2O in $22\,g = \dfrac{22}{18} = 1.22$ mol.

Molar mass of $Fe_2O_3 \cdot xH_2O = 2(55.8) + 3(16.0) + x(18.0)$
$$= (159.6 + 18x) \text{ g mol}^{-1}.$$

Amount of $Fe_2O_3 \cdot xH_2O$ in $100\,g = \dfrac{100}{159.6 + 18x}$ mol.

Amount of H_2O in $100\,g = x \times$ Amount of $Fe_2O_3 \cdot xH_2O$ in $100\,g$

$$= \dfrac{100x}{159.6 + 18x} \text{ mol.}$$

But $\dfrac{100x}{159.6 + 18x} = 1.22 \Rightarrow x = 2.5.$

(d) An underground iron pipe is less likely to corrode if bonded at intervals to magnesium stakes. Give a reason for this. Explain why aluminum would be a poor substitute for magnesium.

Explanation:

Magnesium is above iron in the reactivity series. This means that magnesium would prefer to undergo oxidation than iron. When magnesium is oxidized, the electrons would flow to the iron. This would make the iron less likely to be oxidized.

Although aluminum is also above iron in the reactivity series, the oxidation of aluminum would create a layer of impervious aluminum oxide. As a result, oxygen cannot penetrate this layer of oxide layer. Hence, the aluminum cannot protect the iron from undergoing further oxidation.

CHAPTER 14

AIR AND THE ENVIRONMENT

1. Coal is one of the fossil fuels that is widely used in power stations to generate energy. In the process, it produces sulfur dioxide and oxides of nitrogen. These two gases cause acid rain.

   (a) Nitrogen monoxide, NO, is produced when nitrogen and oxygen react together at a high temperature.

   (i) Write the equation for this reaction.

Explanation:

$$N_2(g) + O_2(g) \rightarrow 2NO(g).$$

Do you know?

— Rainwater is always acidic because the carbon dioxide in the air dissolves in it to form the weak carbonic acid:

$$CO_2(g) + H_2O(l) \rightleftharpoons H_2CO_3(aq).$$

However, if there are additional SO_2 and NO_2 in the air, the acidity of rainwater would increase because of the formation of "extra" acids:

$$SO_2(g) + H_2O(l) \rightleftharpoons H_2SO_3(aq);$$

and

$$4NO_2(g) + O_2(g) + 2H_2O(l) \rightarrow 4HNO_3(aq).$$

(Continued)

(Continued)

This unusually acidic rainwater or other forms of precipitation, such as snow, is known as acid rain. It can have harmful effects on plants, aquatic life, and infrastructure:

- It reacts with metals and carbonates, hence damaging infrastructure.
- It lowers the pH of water bodies, hence affecting aquatic life.
- It reacts with important nutrients from the soil, such as phosphates, hence affecting plant growth.

(ii) Draw the dot-and-cross diagram of nitrogen monoxide.

Explanation:

$$:\overset{\bullet}{N}:\overset{\times}{\underset{\times\times}{O}}\times$$

(iii) Explain why a high temperature is needed for the reaction.

Explanation:

A high temperature is needed to provide energy to overcome the strong N≡N triple bond.

 Q So, does this mean that the activation energy for the above reaction is very high due to the strong N≡N triple bond?

A: You are right!

Do you know?

— When a car or chemical processing plant uses air to burn fuel, the nitrogen from the air is converted to the colorless nitrogen monoxide:

$$N_2(g) + O_2(g) \rightarrow 2NO(g).$$

This nitrogen monoxide can be further oxidized to form the brown nitrogen dioxide:

$$2NO(g) + O_2(g) \rightarrow 2NO_2(g).$$

Nitrogen dioxide is also formed when the heat energy from lightning causes the nitrogen and oxygen to react during a thunderstorm. The smell of nitrogen dioxide can be obvious after the storm. NO_2 dissolves in water to give acid rain and is an irritant to the eyes and lungs.

(iv) Given that the ΔH for the formation of nitrogen monoxide is positive, draw an energy profile diagram for the reaction of nitrogen with oxygen to form nitrogen monoxide.

Explanation:

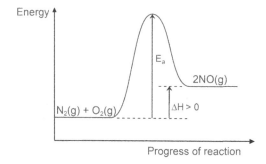

(b) In order to get rid of the sulfur dioxide, many coal-burning power stations are now fitted with a flue-gas-desulfurization plant which removes sulfur dioxide and nitrogen dioxide from the gaseous emissions. In a flue-gas-desulfurization plant, powdered calcium carbonate reacts with sulfur dioxide as shown:

$$SO_2(g) + CaCO_3(s) \rightarrow CaSO_3(s) + CO_2(g).$$

(i) What is the oxidation state of the sulfur atom in sulfur dioxide?

Explanation:

Let the oxidation state of sulfur atom be x.
Therefore, $x + 2(-2) = 0$ {As SO_2 is electrically neutral.}
$$\Rightarrow x = +4.$$

(ii) Given that sulfur dioxide is an acidic gas, what is the nature of the above reaction?

Explanation:

The above reaction is an acid–base reaction since sulfur dioxide is acidic while $CaCO_3$ is basic in nature.

Q Why is CO_3^{2-} basic in nature?

A: The CO_3^{2-} ion comes from the dissociation of the weak carbonic acid, H_2CO_3:

$$H_2CO_3(aq) + 2H_2O(l) \rightleftharpoons CO_3^{2-}(aq) + 2H_3O^+(aq).$$

    acid   base   base   acid

When an acid dissociates, it would form the conjugate base ion. As the CO_3^{2-} ion comes from a weak acid, when it is in water, it undergoes the following basic hydrolysis, resulting in the formation of a basic solution:

$$CO_3^{2-}(aq) + H_2O(l) \rightleftharpoons HCO_3^-(aq) + OH^-(aq).$$

(iii) Suggest why the calcium carbonate is powdered.

Explanation:

Powdered calcium carbonate provides a greater surface area of contact for the SO_2 gas to react. This would minimize the escape of the SO_2 into the atmosphere.

(iv) If nitrogen dioxide behaves similarly to sulfur dioxide, write an equation to show how nitrogen dioxide reacts with calcium carbonate.

Explanation:

$$3NO_2(g) + CaCO_3(s) \rightarrow Ca(NO_3)_2(s) + CO_2(g) + NO(g).$$

(v) State a source of calcium carbonate.

Explanation:

Calcium carbonate comes from limestone.

(vi) Given that nitrogen dioxide forms a dimer easily, draw the dot-and-cross diagram of nitrogen dioxide and use it to explain the ease of dimerization.

Explanation:

Nitrogen dioxide forms a dimer easily because in the dimer, all the atoms would achieve the octet configuration as shown below:

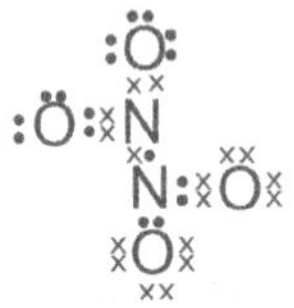

Q Other than using the octet rule, what other explanation can be used to explain why nitrogen dioxide dimerizes?

A: When nitrogen dioxide dimerizes, the energy level of the dimer is lower than that of the monomer because when a bond is formed, energy is evolved. The evolution of energy lowers the energy level of the products. This is the driving force behind the dimerization process.

(vii) What is the color of nitrogen dioxide?

Explanation:

The colour of nitrogen dioxide is brown.

(viii) Calculate the mass of calcium carbonate needed to react with 900 kg of sulfur dioxide.

Explanation:

$$SO_2(g) + CaCO_3(s) \rightarrow CaSO_3(s) + CO_2(g).$$

Molar mass of $SO_2 = 32.1 + 2(16.0) = 64.1 \, g \, mol^{-1}$.

Amount of SO_2 in $900 \, kg = \dfrac{900,000}{64.1} = 1.40 \times 10^4 \, mol.$

Amount of $CaCO_3$ needed = Amount of SO_2 in 900 kg = 1.40×10^4 mol.

Molar mass of $CaCO_3$ = 40.1 + 12.0 + 3(16.0) = 100.1 g mol^{-1}.

Mass of calcium carbonate needed = $1.40 \times 10^4 \times 100.1 = 1.40 \times 10^6$ g.

(c) In the air, sulfur dioxide reacts with nitrogen dioxide forming sulfur trioxide (SO_3). The reactions that take place are shown in the equations:

$$SO_2 + NO_2 \rightarrow SO_3 + NO$$
$$2NO + O_2 \rightarrow 2NO_2.$$

(i) State the oxidation state of the sulfur atom in sulfur trioxide.

Explanation:

Let the oxidation state of sulfur atom be x.

Therefore, $x + 3(-2) = 0$ {As SO_3 is electrically neutral.}
$$\Rightarrow x = +6.$$

(ii) What is the role of nitrogen dioxide in the above reactions? Explain your answer.

Explanation:

The nitrogen dioxide is acting as a catalyst because it remains unchanged at the end of the reaction.

(iii) Sulfur dioxide is used in the Contact Process to make sulfuric acid. Describe the conditions and name the catalyst in the Contact Process.

Explanation:

The temperature used in the Contact Process is about 400°C, with a pressure of about 1 atm, and a ratio of SO_2 to O_2 of 1:1 by volume. The catalyst used is vanadium(V) oxide (V_2O_5).

(d) The carbon dioxide that is emitted during the combustion of fossil fuels is a greenhouse gas. Too much carbon dioxide that is produced through the burning of fossil fuels causes global warming.

   (i) Explain the term *global warming*.

Explanation:

Global warming refers to the additional trapping of heat energy in the Earth's atmosphere brought about by the additional greenhouse gases such as *carbon dioxide*, *methane*, and *nitrous oxide* (N_2O) that are released into the atmosphere due to rampant human activity.

Do you know?

— During the day, the Earth's surface absorbs heat energy emitted from the Sun. And in the night, the Earth cools down by losing this absorbed heat energy. But before all this radiation can escape into outer space, greenhouse gases such as carbon dioxide, water vapor, and others in the atmosphere absorb some of it, which makes the atmosphere warmer. As the atmosphere gets warmer, it makes the Earth's surface warmer too. If not for the greenhouse gases that trap the heat in the atmosphere, the Earth would be a very cold place. This process of keeping the Earth warm is known as the greenhouse effect. Hence, natural greenhouse effect is important to sustain life on Earth.

(ii) Describe two consequences of global warming.

Explanation:

Global warning raises the Earth's temperature, causes polar ice caps to melt, raises the sea level, and eventually submerges cities. In addition, global warming also results in extreme droughts and rainfall around the world, disrupting the growth of plants.

(iii) Explain why carbon dioxide that is produced naturally does not lead to a serious greenhouse effect.

Explanation:

The carbon dioxide that is produced naturally is assimilated by green plants during photosynthesis. Thus, this does not lead to a serious greenhouse effect as long as the carbon cycle is balanced. The following shows the processes in the carbon cycle:

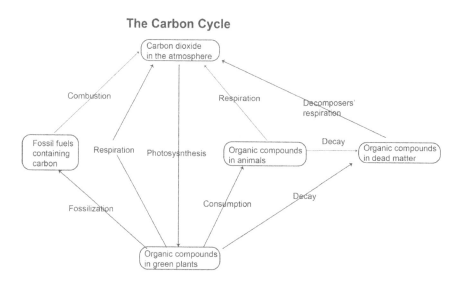

2. Oxides of nitrogen and carbon monoxide are two major air pollutants that are emitted from vehicular exhausts.

   (a) Briefly explain how oxides of nitrogen and carbon monoxide are produced in vehicles.

Explanation:

When car engines use air to burn fuel, oxides of nitrogen are produced due to the reaction of nitrogen from the air with oxygen in the high-temperature car engines.

Carbon monoxide is produced due to the incomplete combustion of the carbon-containing compounds in car engines.

(b) Suggest how oxides of nitrogen and carbon monoxide can be removed from the exhaust gas.

Explanation:

Pollutants that are emitted from the combustion of fuel in car engines can be minimized by passing the exhaust gas through the catalytic converters. These pollutants are converted into less harmful products such as CO_2, N_2, and H_2O.

Do you know?

— The "three-way" catalytic converter consists of a ceramic honeycomb structure coated with the precious metals platinum (Pt), palladium (Pd), and rhodium (Rh), which act as catalysts. A honeycomb structure is used so as to *maximize* the surface area on which the heterogeneously catalyzed reactions can take place. This is because the noble metals used are very expensive.

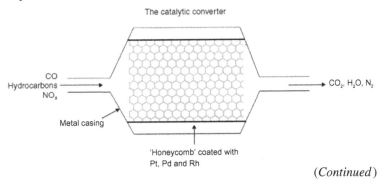

The catalytic converter

CO
Hydrocarbons
NO_x

CO_2, H_2O, N_2

Metal casing

'Honeycomb' coated with
Pt, Pd and Rh

(*Continued*)

(*Continued*)

The chemical reactions that happen in the catalytic converter are as follows:

- As the gases enter, the oxides of nitrogen (NO and NO_2) are reduced to N_2 by the excess CO present, with rhodium acting as the catalyst:

$$2NO + 2CO \xrightarrow{\text{Rh}} N_2 + 2CO_2.$$

- The CO is also oxidized to CO_2 with platinum and palladium as the catalysts:

$$O_2 + 2CO \xrightarrow{\text{Pt/Pd}} 2CO_2.$$

- Non-combustible hydrocarbons are oxidized to CO_2 and H_2O with the help of platinum and palladium.

The catalytic activity of the catalytic converters would be destroyed in the presence of lead. This is because lead is preferentially adsorbed onto the active sites, making it unavailable for the reactants to adsorb on it. Hence, cars that are fitted with catalytic converters must be run on unleaded petrol.

(c) Give chemical equations to show the removal of oxides of nitrogen and carbon monoxide from the exhaust gas.

Explanation:

$$2NO + 2CO \rightarrow N_2 + 2CO_2;$$

and

$$O_2 + 2CO \rightarrow 2CO_2.$$

(d) Explain briefly why carbon monoxide is poisonous.

Explanation:

Carbon monoxide reacts with the hemoglobin in the blood to form carboxyhemoglobin, which prevents red blood cells from taking up oxygen for transport to the rest of the body for respiration.

(e) Why are oxides of nitrogen polluting?

Explanation:

The oxides of nitrogen can dissolve in rainwater and thus increase the acidity of the water. It can also have harmful effects on plants, aquatic life, and infrastructure.

In addition, nitrogen dioxide can also accelerate the break down of ozone. With a thin layer of ozone in the atmosphere, more harmful ultraviolet radiation would reach the Earth's surface. Too much ultraviolet light can damage the skin and cause cancer.

(f) Explain why, in terms of collisions between particles, the rate of production of nitrogen oxide increases as the concentration of oxygen increases.

Explanation:

As the concentration of oxygen increases, the frequency of collisions between the nitrogen and oxygen particles increases. This would lead to a higher frequency of effective collisions; hence, a higher rate of production of nitrogen oxide.

(g) Explain why the rate of the reaction in part (f) increases as the engine temperature increases.

Explanation:

As temperature increases, there are more particles possessing kinetic energies that are sufficient to overcome the activation energy. This thus leads to a higher frequency of effective collisions. In addition, with higher kinetic energies, the particles move faster. Hence, there is a higher frequency of collisions between the particles, which leads to a higher frequency of effective collisions. Thus, the rate of reaction increases.

3. Methane (CH_4) and ozone (O_3) are atmospheric pollutants.

   (a) Briefly describe the sources of methane and ozone in the lower atmosphere.

Explanation:

Methane comes mainly from landfills and decaying organisms. Methane is a potent greenhouse gas.

Ozone in the lower atmosphere is formed by the chemical reactions between oxides of nitrogen and incompletely combusted hydrocarbons. Too much ozone in the lower atmosphere leads to the formation of photochemical smog or haze.

   (b) Explain why methane is a pollutant.

Explanation:

Methane is a greenhouse gas that is capable of trapping heat energy within the atmosphere. It is even more potent than carbon dioxide!

   (c) Ozone is important to help block out the dangerous ultraviolet radiation in the stratosphere. Explain why ozone is a pollutant in the lower atmosphere.

Explanation:

Ozone is a strong oxidizing agent and an irritant of the respiratory system. Thus, in the lower atmosphere, a high level of ozone becomes more of a pollutant.

(d) Describe briefly how ozone is being produced naturally.

Explanation:

Stratospheric ozone is formed naturally when the ultraviolet radiation from the Sun causes the oxygen molecules to react:

$$O_2(g) + UV \text{ light} \rightarrow 2O(g);$$
$$O(g) + O_2(g) \rightarrow O_3(g).$$

Thus, the ozone formation process is in fact an ultraviolet absorption process.

(e) State the cause of ozone depletion.

Explanation:

It has been found that chlorofluorocarbon (CFC) products, which are widely used as aerosol propellant or coolants in refrigerators, catalyze the breaking down of the ozone. This means that the formation of ozone is actually slower than the rate of it being broken down. Hence, there would be a depletion of ozone.

(f) Briefly explain how the compound mentioned in part (e) destroys ozone molecules.

Explanation:

When the CFC molecules reach the stratosphere, it breaks up to give *chlorine atoms* which accelerate the decomposition of ozone. As a result, there is an ozone "hole" in the atmosphere, which allows too much ultraviolet light to reach the Earth's surface.

(g) Draw the dot-and-cross diagrams of both ozone and methane.

Explanation:

$$H_xC_xH \quad \ddot{O}_x\ddot{O}_x\ddot{O}_x$$

(h) At room temperature, ozone decomposes slowly to form oxygen, O_2. The decomposition can be represented by the equation below. The reaction is exothermic. One mole of ozone will release 143 kJ of energy when it is fully decomposed:

$$2O_3(g) \rightarrow 3O_2(g).$$

(i) In terms of the energy changes that take place during bond-breaking and bond-making, explain why this reaction is exothermic.

Explanation:

$$2O-O=O \rightarrow 3O=O.$$

The reaction is exothermic because the energy that is released during the bond-making process is more than the energy that is absorbed during the bond-breaking process.

Q The question did not provide us with the bond energies for the O–O and O=O bonds. So, how are we supposed to know that the breaking of two O–O bonds would absorb less energy than the amount of energy that is released during the formation of a O=O bond?

A: Indeed, the question did not tell contain all those information. But the question did state that the decomposition reaction is exothermic in nature. So, what you are supposed to do is based on the fact that since the reaction is exothermic in nature, it must mean that the bond-making process releases more energy than the energy that is absorbed during the bond-breaking process.

(ii) Draw an energy level diagram for the above reaction.

Explanation:

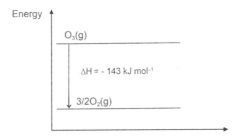

(iii) Explain how you can tell from the energy level diagram that the reaction is an exothermic one.

Explanation:

Since the energy level of the product is lower than that of the reactant, it is an exothermic reaction. This is also indicated by a downward arrow.

(iv) Explain why the rate of this decomposition increases as the temperature increases.

Explanation:

As temperature increases, there are more particles possessing kinetic energies that are sufficient to overcome the activation energy. This thus leads to a higher frequency of effective collisions. In addition, with higher kinetic energies, the particles move faster. Hence, there is a higher frequency of collisions between the particles, which leads to a higher frequency of effective collisions. Thus, the rate of decomposition increases.

(v) Calculate the energy released when 16 g of ozone is decomposed.

Explanation:

Amount of O_3 in 16 g $= \dfrac{16}{48.0} = 0.333$ mol.

One mole of ozone will release 143 kJ when it is fully decomposed, thus the amount of energy that is released by 0.333 mole of ozone $= 0.333 \times 143 = 47.7$ kJ.

(vi) The decomposition of ozone can be accelerated by the presence of a catalyst, the chlorine atom. Explain how a catalyst speeds up the rate of the reaction.

Explanation:

A catalyst lowers the activation energy of the reaction. As a result of the lowered activation energy, there would be more particles possessing kinetic energies that are greater than this lowered activation energy. Hence, the frequency of effective collisions increases, leading to an increase in the rate of the reaction.

CHAPTER 15

ORGANIC CHEMISTRY

1. Crude oil is a mixture of hydrocarbons each with a different melting and boiling point. In a separation process, a sample of crude oil was separated into the following four fractions:

Fraction	Boiling range (°C)
1	room temperature–70
2	70–120
3	120–170
4	170–220

(a) State the separation process that is used to separate the crude oil into its different fractions.

Explanation:

Fractional distillation is used to separate the crude oil into its different fractions.

Do you know?

— The working principle behind fractional distillation of petroleum is that the longer the carbon chain length of the molecule, the higher would its boiling point be. The various components in crude petroleum are heated

(Continued)

(Continued)

and changed into a gas. The gases are passed through a fractional distillation column which becomes *cooler* as the *height increases*. The lower end of the column is *closer to the heat source*; hence, it is hotter.

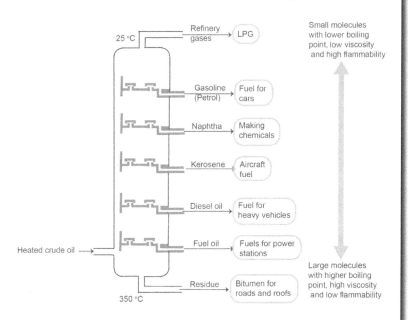

When a compound in the gaseous state cools below its boiling point, it condenses into a liquid. The liquids may be drawn off from the distillation column at various heights. Substances with smaller molecules have lower boiling points and come out at the *top* of the column. Substances with larger molecules have higher boiling points and hence come out at the *bottom* of the column.

(b) What are the two main elements that are present in crude oil?

Explanation:

As crude oil is mainly made up of hydrocarbons, the two main elements that are present in crude oil are hydrogen and carbon.

Do you know?

— The following table shows the different components from crude oil and its uses. From the table, petroleum is mainly used as fuel (LPG, petrol, kerosene, diesel oil) and raw materials in the petrochemical industry.

Components from crude oil	Range of carbon atoms in a molecule	Approximate b.p./°C	Uses
Petroleum gases (mixture of methane, propane, and butane)	1–4	<25	Liquefied petroleum gas (LPG) for cooking and heating.
Petrol/gasoline	5–10	35–75	Fuel for motor vehicles.
Naphtha	7–12	70–170	Petrochemicals to produce plastics, detergents, alcohols, drugs, etc.
Kerosene/paraffin	10–15	170–250	Fuel for aircraft, oil stoves and heating.
Diesel oil	16–20	250–340	Fuel for diesel engines.
Lubricating oil	21–35	340–500	Lubricant in engines; for making waxes and polishes.
Bitumen/asphalt	>70	>500	To pave roads.

(c) Gasoline/petrol is obtained from crude oil. Which one of the four portions in the table most resembles gasoline/petrol?

Explanation:

Fraction 1 in the table most resembles gasoline/petrol. This is because gasoline is used commonly in car engines. Hence, the boiling point cannot be too high otherwise extra heat energy would be needed to vaporize it before the combustion.

> (d) Why does each of the fractions have a boiling range rather than a boiling point?

Explanation:

Each of the fractions has a boiling range rather than a boiling point because each of the fractions is a mixture of compounds with different boiling points. Only a single compound would have a boiling point.

> (e) Crude oil is a non-renewable source for energy. Name two ways in which we can reduce our dependency on crude oil.

Explanation:

The best ways would be to switch to other renewable energy sources such as hydroelectricity, solar energy, wind power, tidal power, geothermal heat, biomass fuel, and the fuel cell. In the meantime, individually, we can help by using as little energy as possible!

> (f) The naphtha fraction is one important fraction obtained from crude oil. State one important use of the naphtha fraction other than as an energy source.

Explanation:

The naphtha fraction is used in the petrochemical industry to produce plastics, detergents, alcohols, drugs, etc.

2. The gas ethene, C_2H_4, is made by the cracking of long-chain hydrocarbon molecules such as $C_{18}H_{38}$.

(a) What is meant by the term *cracking*?

Explanation:

Cracking refers to the process in which large hydrocarbon molecules of heavy heating oil and lubricating oil are broken into smaller molecules such as those in petrol. Cracking is also used to change naphtha into alkenes (especially ethene which is important for polymerization) and to produce hydrogen gas for ammonia synthesis.

Do you know?

— Many kinds of compounds including alkenes are made during the cracking process. Alkenes are formed because there is not enough hydrogen to saturate all bonding positions after the carbon–carbon bonds are broken during the cracking process.

For thermal cracking:		*For catalytic cracking*:	
Reagents:	—	Reagents:	silica-alumina catalyst
Conditions:	800–900°C	Conditions:	450–550°C
Options:	—	Options:	—

— Thermal cracking involves the breaking of strong C–C bonds which requires strong heating. *Homolytic cleavage* of these C–C bonds leads to the formation of free radicals. And just like the free radical substitution reaction, the cracking process is not controllable and a variety of hydrocarbon products is formed depending on the *reaction conditions*.

(*Continued*)

(Continued)

For instance, the cracking of $C_{14}H_{30}$ can give the following products, an alkane and an alkene:

$$C_{14}H_{30} \rightarrow C_8H_{18} + C_6H_{12},$$

or it can break down into more fragments as follows, an alkane and two alkenes:

$$C_{14}H_{30} \rightarrow C_6H_{14} + C_5H_{10} + C_3H_6.$$

Nonetheless, the process is useful in converting long-chain alkanes into shorter-chain alkanes that have more functional usages. The cracking of petroleum oil provides simpler hydrocarbons that can be used as lubricants, fuels, and so forth. Extremely long-chain alkanes are not very volatile and hence it is not easy to make them react in order to get more useful products. In the worst-case scenario where these long-chain alkanes cannot be further broken down, they are used as fuels to heat up boilers in power plants.

— Catalytic cracking is similar to thermal cracking except that lower heating temperatures are needed and a catalyst is used. The catalyst used is none other than the silica-alumina catalyst (SiO_2–Al_2O_3), which is commomly found in clay and sand. After the cracking process is completed, the mixture of products is separated by *fractional distillation*.

Q What is the meaning of homolytic cleavage?

A: When a covalent bond breaks, if the pair of bonding electrons is equally shared between the two separating atoms, resulting in the formation of radicals, this is termed homolytic cleavage. Remember, "homo" means the same. Cleavage is "breaking up." If the pair of bonding electrons is possessed by one of the separating atoms, resulting in the formation of a cation and an anion, we have "heterolytic cleavage."

Q What is a radical?

A: In chemistry, a radical refers to a particle that has unpaired electrons.

Q Why is the temperature for catalytic cracking lower than that for thermal cracking?

A: A catalyst lowers the activation energy of a reaction by providing an alternative pathway for the reaction to take place. With a lowered activation energy, you don't need a higher temperature to make the reaction happen.

(b) Construct an equation to show the cracking of $C_{16}H_{34}$ to make three molecules of ethene and one other product.

Explanation:

$$C_{16}H_{34} \rightarrow 3C_2H_4 + C_{10}H_{22}.$$

(c) Apart from making alkenes and hydrogen, state one other use of cracking long-chain hydrocarbons.

Explanation:

The process is useful in converting long-chain alkanes into shorter-chain alkanes that have more functional usages. The cracking of petroleum oil provides simpler hydrocarbons that can be used as lubricants, fuels, and so forth. Extremely long-chain alkanes are not very volatile and hence it is not easy to make them react in order to get more useful products.

(d) State two uses of the hydrogen gas that is obtained from cracking.

Explanation:

Hydrogen is used as (i) a fuel in big rockets, and in fuel cells; (ii) for the manufacture of ammonia in the Haber Process; and (iii) for the manufacture of margarine from vegetable oil.

(e) The complete combustion of ethene produces carbon dioxide and water.

    (i) Construct a balanced equation for the complete combustion of ethene.

Explanation:

$$C_2H_4(g) + 3O_2(g) \rightarrow 2CO_2(g) + 2H_2O(l).$$

Q It is said that combustion of an alkene is actually a redox process. How can we prove it?

A: If you calculate the oxidation number of the C atoms in ethene, each has a oxidation number of -2. This oxidation number is derived if we consider that the C atom is slightly more electronegative than the H atom. As a result, with two C–H bonds, the C atom would have a -2 oxidation state. Next, the oxidation number of the C atom in CO_2 is $+4$. Therefore, the combustion of alkene to CO_2 is a redox reaction.

Do you know?

— Ethene is the simplest molecule in the alkene family. Alkenes are unsaturated hydrocarbons containing the C=C double bond, which is the functional group in alkene. They have the general formula C_nH_{2n}.

— The melting and boiling points of alkenes increase with an increasing carbon chain length and decrease with an increasing degree of branching for the alkenes. Most alkenes are gases or liquids at room temperature.

(Continued)

(*Continued*)

— Due to the relatively non-polar nature, alkenes are soluble in non-polar organic solvents but are insoluble in polar organic solvents.
— Alkenes, being an *electron-rich* species due to the presence of the C=C double bond, undergo reactions that are mainly *additional* in nature. In this addition reaction, one of the bonds in the C=C double bond is broken.

Q Why does the melting and boiling points increase with an increasing carbon chain length and branching for the alkenes?

A: Alkenes are relatively non-polar molecules. The melting and boiling points of the alkenes only depend on the strength of the instantaneous dipole-induced dipole (id–id) interactions. The longer the carbon chain length, the greater the number of electrons (i.e., the bigger the electron cloud), the more polarizable the electron cloud, and the stronger the id–id interactions.

The higher the degree of branching in the molecule, the smaller the surface area of contact that is possible between the molecules, and the smaller the extent of the id–id interactions.

Q Why does a non-polar solute prefer to dissolve in a non-polar solvent?

A: Energy is needed to overcome the intermolecular forces between the solute molecules and those between the solvent molecules. Dissolution process is just like "digging" a hole between the solvent molecules and then placing the solute molecules within the hole. All these processes need energy. But where does the energy that is needed come from? It has to come from the solute–solvent interactions! Thus, if the solute–solvent interactions are of a similar strength to those of the solute–solute and solvent–solvent interactions, then there would be sufficient amount of energy released from the solute–solvent interactions. This energy would then be able to compensate for the energies that are needed to overcome the solute–solute and solvent–solvent interactions.

Q   What is an addition reaction?

A: An addition reaction is one in which two or more molecules react to give a *single* molecule.

(ii) Calculate the volume of carbon dioxide produced, measured at s.t.p., from the complete combustion of 1.0 g of ethene.

Explanation:

Amount of C_2H_4 in 1.0 g $= \dfrac{1.0}{28.0} = 3.57 \times 10^{-2}$ mol.

Amount of CO_2 produced $= 2 \times$ Amount of C_2H_4 in 1.0 g
$$= 7.14 \times 10^{-2} \text{ mol.}$$

At s.t.p., one mole of gas occupies a volume of 22.7 dm^3.

Hence, volume occupies by the $CO_2 = 7.14 \times 10^{-2} \times 22.7 = 1.6$ dm^3.

(f) Ethene decolorizes bromine dissolved in tetrachloromethane.

(i) State the color change for the reaction.

Explanation:

Bromine dissolved in tetrachloromethane is reddish-brown in color. Thus, the mixture would change from reddish-brown to colorless.

Do you know?

— The bromination of alkene serves as a useful distinguishing test to determine the presence of an alkene in an unknown organic sample. If you are provided with two test-tubes, one of which contains an alkene and the other an alkane, the distinguishing test using Br_2 must be carried

(Continued)

(Continued)

out in the dark. This is done so as to prevent the substitution of alkane from occurring in the presence of light. There will then be one positive result — the decolorization of the reddish-brown Br_2 for the test-tube that contains the alkene; and a negative result — no decolorization of the reddish-brwn Br_2 for the test-tube that contains the alkane.

(ii) Name the above reaction.

Explanation:

The above reaction is an addition of bromine reaction.

(iii) Draw the dot-and-cross diagrams for ethene and the product formed after reacting with bromine.

Explanation:

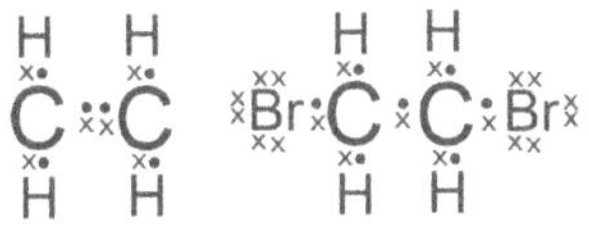

(iv) Ethene is a gas at room temperature while the product that is formed is a liquid at room temperature. Explain why there is a difference in the two physical states.

Explanation:

Ethene is a non-polar molecule with id–id interactions between the molecules. The product (1,2-dibromoethane) is a polar molecule with permanent dipole–permanent dipole (pd–pd) interactions. Since the pd–pd interactions are stronger than the id–id interactions, ethene is a gas while the product that is formed is a liquid at room temperature.

> (g) Ethene reacts with hydrogen gas in the presence of a catalyst to give a saturated organic compound.
>
> (i) State the name of the reaction.

Explanation:

The reaction is an addition of hydrogen reaction, also known as catalytic hydrogenation.

Do you know?

— The hydrogenation of alkene gives an alkane. Catalytic hydrogenation is an example of what is called *heterogeneous* catalysis whereby the catalyst and the reactants are in *different phases*. Commercial applications include the use of powdered nickel in the hydrogenation of unsaturated fats to the saturated form in the manufacture of margarine:

$$R-CH=CH-R'(l) + H_2(g) \rightarrow R-CH_2-CH_2-R'(l).$$

Q   What is the purpose of hydrogenating unsaturated fats to saturated fats?

A: The presence of C=C bonds make unsaturated fats more prone to free radical attacks at the electron-rich sites. This easily makes the fats rancid (turn sour as carboxylic acids are formed) and thus have a shorter shelf-life in the market. But take note that saturated fat has a higher melting point than unsaturated fats, which is not very good for our body.

 Why is the melting point of unsaturated fats lower than that of saturated fats?

A: Due to the presence of multiple C=C bonds along the carbon skeleton of an *unsaturated* fat molecule, the three-dimensional structure of the molecule is relatively more *spherical* in shape. In contrast, the molecule of a *saturated* fat is more *linear*. As a result, the surface area of contact for the saturated fat molecule is greater than that for the unsaturated fat molecule. Hence, more extensive intermolecular forces of attraction account for the higher melting point of saturated fats. This would mean that at the *physiological* body temperature, saturated fat molecules would more likely solidify as compared to unsaturated fat molecules. The solidification would be more likely to cause arteriole blockage.

(ii) Write down a balanced equation for the reaction.

Explanation:

$$C_2H_4(g) + H_2(g) \rightarrow C_2H_6(g).$$

(iii) Suggest a simple chemical test to differentiate between ethene and the organic product formed.

Explanation:

Bromine dissolved in tetrachloromethane can be used as a distinguishing test. Ethene would decolorize the reddish-brown bromine, while the organic product (alkane) formed will not.

(iv) One mole of the organic product formed reacts with one mole of chlorine molecules. Construct a balanced equation for this reaction.

Explanation:

$$C_2H_6(g) + Cl_2(g) \rightarrow C_2H_5Cl(g) + HCl(g).$$

Q Why is the reaction equation not $C_2H_6(g) + Cl_2(g) \rightarrow C_2H_4Cl_2(g) + H_2(g)$?

A: This has to do with the mechanism of the reaction. If you are interested, you can refer to *Understanding Advanced Organic and Analytical Chemistry* by K.S. Chan and J. Tan.

Do you know?

— Alkanes, although unreactive with seemingly few reactions, do undergo substitution with halogens (Cl_2 or Br_2), but only in the presence of stringent conditions such as *strong heating* or *ultraviolet light*. Each hydrogen atom in ethane can be substituted by a bromine atom.

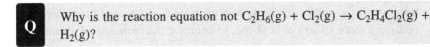

If all the hydrogen atoms are replaced, the product formed is as follows:

$$
\begin{array}{c}
\text{H} \quad \text{H} \\
| \quad | \\
\text{H—C—C—H} \\
| \quad | \\
\text{H} \quad \text{H}
\end{array}
\longrightarrow
\begin{array}{c}
\text{Br} \quad \text{Br} \\
| \quad | \\
\text{Br—C—C—Br} \\
| \quad | \\
\text{Br} \quad \text{Br}
\end{array}
$$

(v) Name the catalyst that is used in this reaction.

Explanation:

The catalyst used is powdered nickel catalyst.

(vi) State one important industrial application of this reaction.

Explanation:

In the manufacture of margarine, unsaturated fats are hydrogenated to the saturated form.

(vii) Explain how the catalyst speeds up the rate of the reaction.

Explanation:

A catalyst lowers the activation energy of the reaction by providing an alternative pathway for the reaction to take place. As a result of the lowered activation energy, there would be more particles possessing kinetic energies that are greater than this lowered activation energy. Hence, the frequency of effective collisions increases, leading to an increase in the rate of the reaction.

3. Crude oil contains mainly alkanes. The first three members of the alkane family are methane, ethane, and propane. They are all gases at room temperature.

   (a) Rank methane, ethane, and propane in order of decreasing boiling point. Explain the trend.

Explanation:

In order of decreasing boiling point: propane > ethane > methane.

The melting and boiling points of the alkanes only depend on the strength of the id–id interactions. The longer the carbon chain length, the greater the number of electrons (i.e., the bigger the electron cloud), the more polarizable the electron cloud, and the stronger the id–id interactions. Hence, this accounts for the higher boiling point of propane compared to ethane, and ethane compared methane.

(b) Which one of the alkanes is the main component of natural gas?

Explanation:

Methane is the main component of natural gas.

(c) Natural gas can be used as a domestic fuel for cooking. A substance with a characteristic smell is added before it is supplied to consumers. What is the reason for giving natural gas a characteristic smell?

Explanation:

Natural gas is colorless and odorless. Thus, if there is a gas leak, it is dangerous as the gas is flammable. Hence, mercaptan, a non-toxic gas, is added to give natural gas a characteristic smell.

Q   What is mercaptan?

A: It is a flammable gas with the formula CH_3SH, which is similar to methanol (CH_3OH). Mercaptan can be found in our blood, brain, plant tissues, animal feces, nuts, cheese, and bad breath.

(d) The complete combustion of methane produces carbon dioxide and water.

    (i) Construct a balanced equation for the complete combustion of methane.

Explanation:

$$CH_4(g) + 2O_2(g) \rightarrow CO_2(g) + 2H_2O(l).$$

 Q Why is combustion of an alkane exothermic? In addition, why is the combustion of the alkane more exothermic with more carbon atoms that are present in an alkane molecule?

A: You can use the data from bond energies to prove this. Essentially, the energy that is released during the formation of the strong bonds in CO_2 and H_2O is much more than the energy that is required to break the C–H, C–C and O=O bonds. Hence, you have an exothermic reaction. And with more carbon atoms present in an alkane molecule, more CO_2 and H_2O are formed. Thus, more exothermic will be the combustion process.

    (ii) Under a limited supply of air, methane burns with a smoky yellow flame. Name *two* other possible products formed when methane is burned in a limited supply of air.

Explanation:

If the supply of oxygen is limited, incomplete combustion occurs and products such as non-combusted carbon (in the form of soot) and CO are formed, apart from the CO_2 and H_2O, i.e.,

Incomplete combustion:

$$CH_4(g) + \frac{3}{2}O_2(g) \rightarrow CO(g) + 2H_2O(l);$$

and

$$CH_4(g) + O_2(g) \rightarrow C(s) + 2H_2O(l).$$

(iii) Calculate the volume of carbon dioxide produced when $20\,cm^3$ of methane is completely burned in excess oxygen.

Explanation:

$$CH_4(g) + 2O_2(g) \rightarrow CO_2(g) + 2H_2O(l).$$

According to Avogadro's Law, at the same temperature and pressure, the volume of a gas is directly proportional to the number of moles, i.e., $V \propto n$. Hence, $20\,cm^3$ of CH_4 will react with $40\,cm^3$ of O_2 to give $20\,cm^3$ of CO_2.

(e) Cracking is the breaking up of large alkane molecules into smaller molecules. Name the two types of non-alkane molecules that may be formed by cracking.

Explanation:

Hydrogen and alkenes may be formed by cracking.

(f) Methane reacts with chlorine gas.

　(i) State the necessary condition for the above reaction to take place.

Explanation:

The presence of ultraviolet light or heat is important for the reaction to take place.

　(ii) Name the above reaction.

Explanation:

The reaction is a substitution reaction. (To be more specific, it is a free radical substitution reaction because the reaction is initiated by a free radical.)

Do you know?

— There are no electrostatically attractive features on an alkane molecule that draw in a potential attacker species. C–H bonds are essentially non-polar as both the carbon and hydrogen atoms have a rather small electronegativity difference. Thus, alkanes are relatively chemically inert as both the carbon and hydrogen atoms in an alkane molecule are neither electron rich nor electron deficient. They undergo *few* reactions, mainly combustion and *substitution* reactions.

(iii) Draw the dot-and-cross diagram of the product that is formed under the excess chlorine condition.

Explanation:

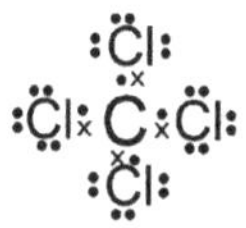

(iv) Write down a balanced equation for the reaction in part (f)(iii).

Explanation:

$$CH_4(g) + 4Cl_2(g) \rightarrow CCl_4(l) + 4HCl(g).$$

4. Olive oil contains oleic acid which is unsaturated.

   (a) Give a chemical test that would show that oleic acid is an acid. State the observation and name the chemical reaction that takes place in the chemical test.

Explanation:

To show that oleic acid is an acid, we can add some aqueous Na_2CO_3 to it. If an effervescence of CO_2 is observed and the gas evolved gives a white precipitate with aqueous $Ca(OH)_2$, oleic acid is acidic in nature.

The chemical reaction is an acid-base reaction.

Q   Can we use sodium metal as a test to show that oleic acid is acidic?

A: Yes, you can. If an effervescence of H_2 gas is observed and the gas extinguished a lighted splint with a "pop" sound, it is a confirmation that oleic acid is acidic.

Do you know?

— Carboxylic acids are organic compounds that contain the following carboxyl functional group, denotes as –COOH.

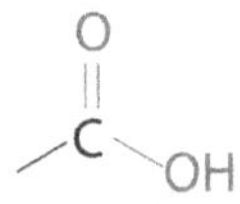

$$\begin{array}{c} O \\ \parallel \\ \diagup C \diagdown \\ OH \end{array}$$

From the constitutional/structural formula, the carboxyl functional group contains a hydroxyl group. Thus, we would expect a carboxylic acid to have the same properties as those of an alcohol. The general formula of a carboxylic acid is $C_nH_{2n+1}COOH$, where $n = 0$ for the first member, methanoic acid (HCOOH).

— Other than carboxylic acids being acidic in nature, alcohols are acidic too! But carboxylic acids are *stronger* acids as compared to alcohols.

(Continued)

(*Continued*)

This can be observed from their behavior toward bases. Alcohols cannot react with a weak base!

$$2CH_3CH_2OH(aq) + 2Na(s) \rightarrow 2CH_3CH_2ONa(aq) + H_2(g);$$

$$2CH_3COOH(aq) + 2Na(s) \rightarrow 2CH_3COONa(aq) + H_2(g);$$

$$CH_3COOH(aq) + NaOH(aq) \rightarrow CH_3COONa(aq) + H_2O(l);$$

$$CH_3COOH(aq) + NaHCO_3(aq) \rightarrow CH_3COONa(aq) + H_2O(l) + CO_2(g);$$

and

$$2CH_3COOH(aq) + Na_2CO_3(aq) \rightarrow 2CH_3COONa(aq) + H_2O(l) + CO_2(g).$$

In the above reactions, the corresponding salt of the carboxylic acid is formed, e.g., methanoic acid forms the salt methanoate, ethanoic acid forms ethanoate, propanoic acid forms propanoate, etc.

Q Why is the carboxylic acid a stronger acid than its corresponding alcohol of similar carbon chain length?

A: There is an additional highly electronegative oxygen atom in the carboxylic acid functional group, –COOH, which is absent in the alcohol functional group, –OH. This causes the O–H bond in the carboxylic acid functional group to be much weaker than that in alcohol. Hence, carboxylic acid is a better Brønsted–Lowry acid, i.e., a better proton donor.

Q Since sodium reacts with both alcohols and carboxylic acids, sodium is too reactive to be able to differentiate these two compounds. But since NaOH, NaHCO$_3$, and Na$_2$CO$_3$ can react with carboxylic acids but not with alcohols, can we use these three compounds to differentiate an alcohol from a carboxylic acid?

A: Yes, you can! This shows that you need to use something that is weaker or less reactive to differentiate the strengths of two "relatively" strong acids. Well, the strength of an acid or a base, or the reactivity of a metal or oxidizing power or reducing power, etc., are all very relative matters!

(b) Give a chemical test that would show that oleic acid is unsaturated. State the observation and name the chemical reaction that takes place in the chemical test.

Explanation:

Bromine dissolved in tetrachloromethane can be used as a chemical test. Oleic acid would decolorize the reddish-brown bromine as it is unsaturated, containing the C=C double bonds.

The name of the chemical reaction is addition reaction.

(c) The molecular formula of oleic acid is $C_{17}H_{31}CO_2H$.

(i) Explain the term *molecular formula*.

Explanation:

The molecular formula of a compound is a representation of the actual whole number ratio of the different atoms that made up the compound.

(ii) State the number of double bonds present in oleic acid. List the type of double bonds.

Explanation:

There are three double bonds in $C_{17}H_{31}CO_2H$. There are two C=C double bonds and one C=O double bond belonging to the –COOH functional group.

Q How do you determine the number of double bonds in an organic compound?

A: Simple! Count the carbon atoms first. For n carbon atoms, if it is an alkane, then there should be $(2n + 2)$ of H atoms present. For example, oleic acid has a total of 18 carbon atoms, so there should be 38 H atoms if there are no double bonds in the molecule. Now, since there are only 32 H atoms, there should be three double bonds. This is because one double bond is equivalent to two H atoms!

(iii) If all the 17 carbon atoms of oleic acid formed a linear chain, list down three possible structures of oleic acid.

Explanation:

The three possible structures are:

$$CH_3CH=CHCH=CH(CH_2)_{12}COOH;$$

$$CH_3CH_2CH=CHCH=CH(CH_2)_{11}COOH; \quad \text{and}$$

$$CH_3(CH_2)_3CH=CH(CH_2)_2CH=CH(CH_2)_7COOH.$$

(*Note*: There are many more possible structures other than these three that are listed here!)

(iv) What is the relationship between the molecules in part (c)(iii)?

Explanation:

The molecules are constitutional/structural isomers of each other.

Do you know?

— Compounds that have the same molecular formula but different structures are known as *isomers*. This phenomenon is known as *isomerism*. Isomers generally have different physical and chemical properties, but they can also have *similar* chemical properties if they contain the *same* functional groups. Each specific functional group possesses a characteristic set of chemical reactions.

— Constitutional/structural isomers are compounds with the same *molecular formula* but different structures or *constitutional/structural formulae*. Both butene and cyclobutane constitute a pair of constitutional/structural isomers.

Skeletal formula   Displayed formula

The difference in structures can be attributed to either a difference in the arrangement of atoms or due to the presence of different functional groups.

Based on the above definitions, constitutional/structural isomerism can be classified into three main types:

- *chain* isomerism;
- *positional* isomerism; and
- *functional* group isomerism.

But at this level, we are going to discuss only about chain isomerism.

— Compounds that exhibit chain isomerism with each other have the *same* functional group but differ in the way the carbon atoms are connected in the *main skeletal* carbon chain of their molecules. In other words, these molecules differ in the *degree of branching*, hence the term *chain isomers*. The *lengths* of the *main* carbon skeletal structure for chain isomers are *not* the same. An example would be pentane, dimethylpropane, and methylbutane:

Pentane   Dimethylpropane   Methylbutane

Q   Do chain isomers have the same chemical properties?

A: Yes, they do. This is because chain isomers contain the *same functional group* and hence have the same chemical properties.

(v) Suggest a physical property that is different for the three molecules.

Explanation:

The three molecules would have different boiling and melting points.

(vi) If the three molecules form a mixture, suggest how you would separate them physically.

Explanation:

Since the three molecules have different boiling points, they can be separated via fractional distillation.

(d) When oleic acid is heated with ethanol in the presence of a substance **A**, a sweet-smelling compound **B** is formed.

   (i) Suggest an identity for **A** and **B**.

Explanation:

Compound **A** is concentrated H_2SO_4, while compound **B** is an ester.

Do you know?

— Alcohols react with carboxylic acids to form esters. The esterification process is a reversible reaction and it occurs at such a slow rate that equilibrium is attained only after a few hours. As such, concentrated H_2SO_4 or other acids, is used as a catalyst to speed up the reaction.

The name of an ester consists of two parts, for example, ethyl methanoate:

"Ethyl" comes from ethanol while the "methanoate" comes from methanoic acid.

Q   Why is the esterification reaction also termed as "condensation reaction"?

A: Take note that in organic chemistry, when two molecules come together to form another molecule through covalent bonding, and at the same time, a small molecule such as water, ammonia, or hydrochloric acid is expelled, the reaction can be termed *condensation* reaction. Please do not be confused with the other condensation process, which refers to the conversion of a gaseous substance to its liquid state.

(ii) Suggest a simple chemical test to differentiate between ethanol and compound **B**.

Explanation:

Add a piece of sodium metal into both ethanol and compound **B**. Only ethanol would give an effervescence of H_2 gas. The H_2 gas would extinguish a lighted splint with a "pop" sound.

Do you know?

— Alcohol dissolves in water to give a neutral solution although it has the acidic −OH group. In fact, it is a weak acid that is able to react with a reactive metal to give hydrogen gas:

$$H-\underset{\underset{H}{|}}{\overset{\overset{H}{|}}{C}}-\underset{\underset{H}{|}}{\overset{\overset{H}{|}}{C}}-O-H \ + \ Na \ \longrightarrow \ H-\underset{\underset{H}{|}}{\overset{\overset{H}{|}}{C}}-\underset{\underset{H}{|}}{\overset{\overset{H}{|}}{C}}-O^-Na^+ \ + \ \tfrac{1}{2}H_2$$

One can thus determine for the presence of the alcohol by testing the H_2 gas that is evolved with a lighted splint, a "pop" sound would be heard.

Q Does that mean that sodium metal is too reactive to be able to differentiate the strengths of the two acids, water and alcohol, since both of them can react with Na to give hydrogen gas?

A: Yes, you are right! So, in order to differentiate these two weak acids, you probably need to use a less reactive metal such as magnesium.

Q Other than the hydrogen gas, is the other compound a salt?

A: Yes, it is in fact an *organic salt*. It is ionic in nature too! So, you see, a salt is not necessary an inorganic product that is generated from what we have covered previously.

(iii) Suggest a simple chemical test to differentiate between oleic acid and compound **B**.

Explanation:

Add a piece of sodium metal into both oleic acid and compound **B**. Only oleic acid would give an effervescence of H_2 gas. The H_2 gas would extinguish a lighted splint with a "pop" sound.

> (iv) Suggest a simple chemical test to differentiate between ethanol and oleic acid.

Explanation:

Add aqueous Na_2CO_3 into both oleic acid and ethanol. Only oleic acid would give an effervescence of CO_2 gas. The CO_2 gas would give a white precipitate with aqueous $Ca(OH)_2$.

Do you know?

— Some uses of ethanol:

- Ethanol is a polar solvent which can dissolve many substances that water cannot. Due to its higher volatility, it is widely used as a medium for perfumes, paint, deodorants, etc.
- Ethanol can be used as a fuel by mixing it with petrol.
- Ethanol is a good antiseptic as it can denature enzymes in bacteria.
- Ethanol can be drunk and it has an intoxicating effect. Too much ethanol damages the liver and kidneys, and causes addiction.

Q   Why is ethanol so soluble in water?

A: The ability to form hydrogen bonds accounts for the high solubility of short-chain alcohols in water. However, as the length of the carbon chain increases, the solubility of the long-chain alcohol in water decreases due to the increasing size of the *hydrophobic* alkyl chain. Yet, it is the presence of the hydrophobic alkyl chain that renders alcohols also miscible with most organic compounds.

> **Q** What is the meaning of hydrophobic?

A: "Hydro" means water, "phobic" comes from phobia which means "dislike." Thus, hydrophobic means "disliking water" or "incompatible with water." The opposite is hydrophilic, in which "philic" means "love."

5. Ethanol can be made from glucose ($C_6H_{12}O_6$) by fermentation.

   (a) Draw the dot-and-cross diagram of ethanol.

Explanation:

Do you know?

— Industrially, ethanol can be prepared by passing a mixture of ethene and steam over phosphoric(V) acid (catalyst) at a high temperature of 300°C and a high pressure of 60 atm:

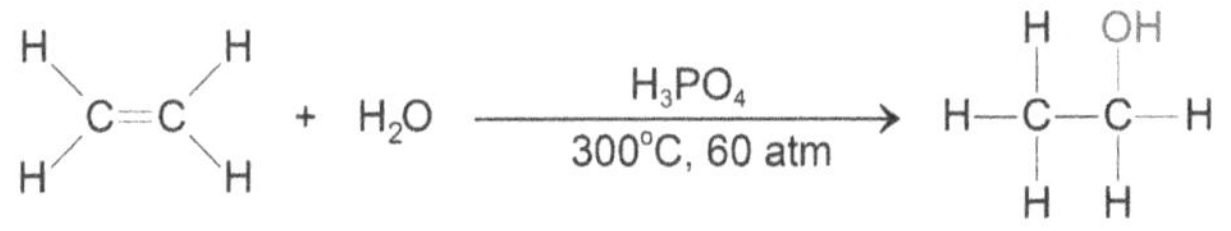

The ethanol produced is not drinkable due to the presence of impurities. Ethanol can be dehydrated to form the alkene in the presence of concentrated sulfuric acid. This is actually the reverse of the hydration of an alkene to form an alcohol.

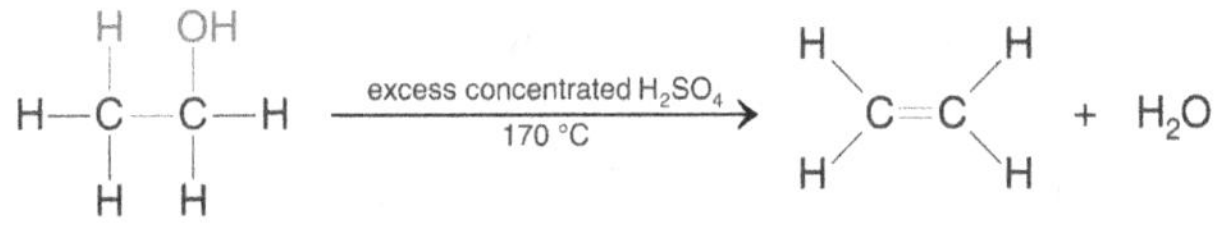

— Drinkable ethanol can be produced by the fermentation of carbohydrates using microorganisms such as yeast, which contains biological catalyst called enzymes. The carbohydrates, such as sugar, are broken down to *ethanol* and *carbon dioxide*. The process is carried out in an oxygen-free environment as the reaction is *anaerobic*:

$$C_6H_{12}O_6(aq) \rightarrow 2CH_3CH_2OH(aq) + 2CO_2(g).$$

Q Why must the fermentation be anaerobic?

A: This is because the bacteria in the air would further oxidize the ethanol into ethanoic acid (CH_3COOH), which makes the solution sour:

$$CH_3CH_2OH(aq) + O_2(g) \rightarrow CH_3COOH(aq) + H_2O(l).$$

Ethanoic acid is the molecule in vinegar that causes it to taste sour. In fact, vinegar is produced by further fermentation of ethanol.

Do you know?

— Like other non-biological catalysts, enzymes catalyze reactions by providing an *alternative* reaction pathway with *lower* activation energy. This occurs when the enzyme forms a complex with the substrate or substrates (reactants) of the reaction. Thus, a simple picture of the enzyme action is:

substrate + enzyme → enzyme/substrate complex → enzyme + products

(b) Write the equation for the preparation of ethanol by this method.

Explanation:

$$C_6H_{12}O_6(aq) \rightarrow 2CH_3CH_2OH(aq) + 2CO_2(g).$$

Do you know?

— Ethanol that is produced from the fermentation of sugar can be used as a fuel, replacing petrol. It is less volatile, which makes it less flammable and safer to handle.

(c) Suggest a chemical test for the gas that is evolved during the fermentation process.

Explanation:

Bubble the gas into aqueous $Ca(OH)_2$; a white precipitate of $CaCO_3$ will form if the gas is CO_2.

> (d) What other substance could be added to a solution of glucose in order to make ethanol?

Explanation:

Microorganisms such as yeast, which contains biological catalyst called enzymes, could be added.

> (e) Explain why the control of temperature is important in the fermentation process.

Explanation:

Higher temperatures may denature the enzymes. Thus, the control of temperature is important.

Do you know?

— Enzymes, which consist of proteins or polypeptides, are an important class of biological catalysts, speeding up chemical reactions in living systems. Without them, most biochemical reactions would be too slow to sustain life. Examples of enzymes include amylase (for hydrolyzing starch) and lipase (for hydrolyzing fats and lipids). Enzymes possess the following characteristics:

1. *Nature and size*
   Enzymes are *globular* proteins which fold in a particular conformation, creating a three-dimensional *active site*. They have relative molecular masses of around 10^5 to 10^7.

(Continued)

(*Continued*)

2. *Efficiency*

Enzymes are required in very small amounts; they are very *effective* catalysts. This is because the enzyme molecules are regenerated after catalysing a reaction. A typical enzyme molecule may be regenerated a million times in one minute.

3. *Specificity*

Due to the special conformation of the three-dimensional active site in which only certain molecules can be adsorbed, enzymes are very *specific* to a particular reaction or type of reaction.

4. *Temperature*

A high temperature of above 50°C results in a *change* of the three-dimensional active site; this *denatures* the enzyme, rendering it inactive. Thus, enzymes operate most effectively at body temperature, of about 37°C.

5. *Sensitivity to pH*

A pH change results in a *change* of the three-dimensional active site; this *denatures* the enzyme, rendering it inactive. Different enzymes have different optimal pH levels.

Q   What is the meaning of "adsorb"?

A: Adsorb refers to the phenomenon whereby the particle "sticks" onto the surface. Of course, it is the electrostatic attractive force that "bonds" the particle to the surface. Hence, adsorption is an interfacial phenomenon. As for absorption, the particle would have entered into the "body" of the object, not unlike a sponge absorbing water.

(f) What is the mass of ethanol that could be obtained from 30 g of glucose, assuming there is only 40% conversion of glucose into ethanol?

Explanation:

Molar mass of $C_6H_{12}O_6 = 6(12.0) + 12(1.0) + 6(16.0) = 180$ g mol^{-1}.

Amount of $C_6H_{12}O_6$ in 30 g = $\frac{30}{180}$ = 0.167 mol.

If there is only 40% conversion, amount of $C_6H_{12}O_6$ converted = $\frac{40}{100} \times 0.167 = 6.67 \times 10^{-2}$ mol.

Amount of $CH_3CH_2OH = 2 \times 6.67 \times 10^{-2} = 0.133$ mol.

Mass of ethanol obtained = $0.133 \times 46 = 6.13$ g.

(g) Name the method used for isolating ethanol from the mixture of ethanol and glucose solution.

Explanation:

The method used is distillation.

(h) Explain why the method mentioned in part (g) works.

Explanation:

The boiling point of ethanol is much lower than that of the glucose solution. Thus, ethanol can be effectively separated from the glucose solution through simple distillation.

 Q Why is the boiling point of ethanol (78.4°C) so much different from that of water?

A: A water molecule, on the average, is capable of forming two hydrogen bonds per water molecule. Replacing one of the hydrogen atoms of a water molecule by a carbon atom, as in an alcohol, results in only one hydrogen bond that is possible per alcohol molecule. Hence, the hydrogen bonding in ethanol is less extensive than that in water.

(i) When concentrated sulfuric acid is added to ethanol and heated to about 170°C, a colorless flammable gas is evolved.

   (i) Give the chemical equation for the above reaction.

Explanation:

$$CH_3CH_2OH(l) \rightarrow CH_2{=}CH_2(g) + H_2O(l).$$

(ii) Name the above reaction.

Explanation:

The above reaction is a dehydration reaction.

(iii) State the role of the concentrated sulfuric acid.

Explanation:

The concentrated sulfuric acid is acting as a catalyst for the dehydration reaction.

Q   If we use dilute sulfuric acid, would the dehydration still work?

A: No! For further details, refer to *Understanding Advanced Organic and Analytical Chemistry* by K.S. Chan and J. Tan.

(iv) State the functional group that is present in the gas and suggest a simple chemical test for the functional group.

Explanation:

The functional group present in the gas is a C=C double bond. Bromine dissolved in tetrachloromethane can be used as a chemical test. The gas containing the C=C double bond would decolorize the reddish-brown bromine as it is unsaturated. The chemical reaction is an addition reaction.

> (v) State one important use of the gas other than to convert it to ethanol or combust it to get energy.

Explanation:

The ethene gas can be used to make poly(ethene), which is a polymer used to make plastic film.

Do you know?

— Polymers are macromolecules that are made up of many smaller molecules combined to form very long chains. The number of molecules used to form a polymeric molecule can easily be in the hundreds of thousands or even millions.

— Not all molecules can come together to form polymers. Those that can do so are termed *monomers* and a polymer is made up of many *repeating units* of monomers. When subjected to the polymerization process, new covalent bonds are formed between the monomers, linking them up to form either long linear chains or an extensive three-dimensional structural network.

— Monomers that contain unsaturated carbon–carbon double (C=C) or triple bonds (C≡C) bind together to form an *addition polymer*. When these monomers combine to form the polymer, there is *no loss* of any segment of the monomeric unit, thus giving rise to the term "*addition*" polymer. Obviously, such a polymerization process is known as *addition polymerization*.

— The following table shows some common examples of addition polymers and their properties.

(Continued)

(*Continued*)

Monomer	Polymer	Properties	Examples of uses
Propene CH_3, H ... $C=C$ ... H, H	Poly(propene) $+CH_2-CH+$ ... CH_3 ... n	— Tough and flexible — Resistant to chemicals — Resistant to heat	— Packaging — Ropes — Polymer banknotes — Stationery
Chloroethene (vinyl chloride) Cl, H ... $C=C$ ... H, H	Poly(chloroethene) or poly(vinyl chloride) $+CH_2-CH+$ ... Cl ... n	— Resistant to chemicals — Relatively unstable to heat and light	— Flexible hoses — Electrical cable insulation — Pipes — Waterproof clothing — Shoes and bags — Signage
Phenylethene (styrene) $CH_2=CH-$ ⬡	Poly(phenylethene) or poly(styrene) $+CH_2-CH+$ ... ⬡ ... n	— Resistant to chemicals — Hard with limited flexibility — Transparent	— Disposable cutlery — Foam drink cups — CD cases — Insulation materials
Tetrafluoroethene F, F ... $C=C$ ... F, F	Poly(tetrafluoroethene) (brand name Teflon) $+CF_2-CF_2+$ ... n	Its properties stemmed from the aggregate effect of strong C–F bonds: e.g., — Resistant to heat and chemicals — "Anti-stick" properties — Low friction	— Non-stick coating for cookware — Lubricant — Laboratory apparatus

6. Ethanoic acid is made from ethanol via oxidation.

   (a) Give the full constitutional/structural formulae of ethanoic acid and ethanol.

Explanation:

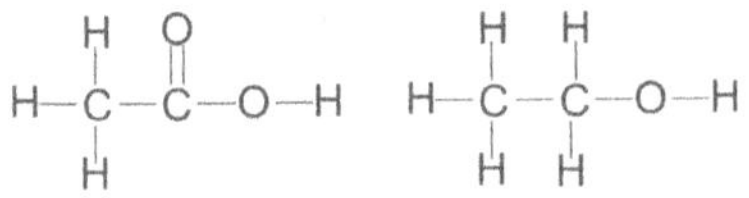

   (b) Construct a balanced equation for the above reaction.

Explanation:

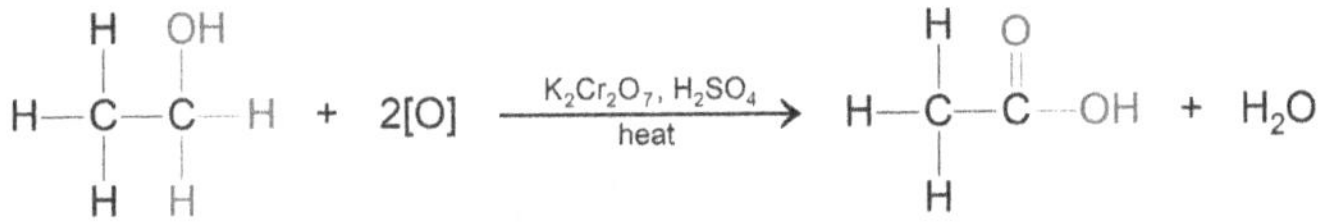

Q   What is the meaning of [O]?

A: In organic chemistry, we normally use the "[O]" symbol to represent the addition of oxygen atoms in an oxidation reaction in order to balance the chemical equation, if the oxidizing agents used are either acidified $KMnO_4$ or acidified $K_2Cr_2O_7$.

Do you know?

— Industrially, ethanoic acid can be prepared by oxidizing ethanol in the presence of hot acidified potassium dichromate(VI). The ethanoic acid produced is not drinkable due to the presence of impurities. Drinkable ethanoic acid can be produced by the fermentation of carbohydrates using microorganisms.

(c) State the reagents and conditions used for the conversion of ethanol to ethanoic acid.

Explanation:

The reagents are $K_2Cr_2O_7$(aq) acidified with H_2SO_4(aq) and heating conditions are needed.

(d) Name a common substance used at home that contains ethanoic acid.

Explanation:

Ethanoic acid is found in vinegar.

Q Why is ethanoic acid so soluble in water to form vinegar?

A: The ability to form hydrogen bonds accounts for the high solubility of the short-chain carboxylic acids in water. However, as the length of the carbon-chain increases, their solubility in water will decrease due to the increasing size of the hydrophobic alkyl chain. The latter is responsible for carboxylic acid's increasing solubility in non-polar solvents.

(e) How would you separate a mixture of ethanol and ethanoic acid?

Explanation:

Since ethanol and ethanoic acid have a great difference in their boiling points, they may be separated via simple distillation.

> **Q** How do you then account for the higher boiling point of ethanoic acid, CH_3COOH, compared to that of ethanol CH_3CH_2OH?

A: There are two factors here: (1) the number of hydrogen bonding sites in a –COOH group (four lone pairs and a H atom) is more than those in a –OH (two lone pairs and a H atom) group, and hence, the hydrogen bonding in carboxylic acid is more *extensive* than that in an alcohol of similar molecular weight; and (2) the hydrogen bond that is formed between the H atom of a carboxylic acid is stronger than that for an alcohol as the H atom for the carboxylic acid is more electron deficient due to the presence of another highly electronegative O atom in the –COOH group. This additional highly electronegative O atom "helps" to pull the electron density away from the O–H bond of the –COOH group, thus making the H atom more electron deficient.

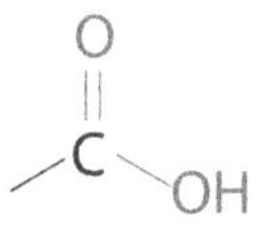

(f) Ethanol and ethanoic acid react together to produce another organic compound.

(i) Give a name for the above reaction.

Explanation:

The reaction is esterification or condensation reaction.

(ii) Give the full constitutional/structural formula for the organic compound formed.

Explanation:

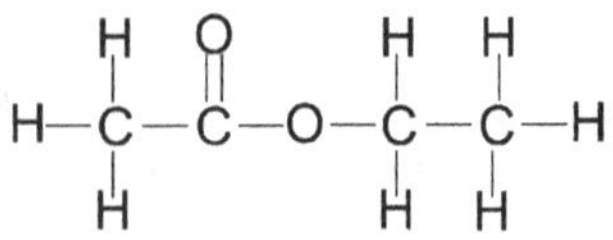

(iii) Construct a balanced equation for the above reaction.

Explanation:

$$H-\overset{\overset{\displaystyle H}{|}}{\underset{\underset{\displaystyle H}{|}}{C}}-\overset{\overset{\displaystyle O}{\|}}{C}-O-H \; + \; H-\overset{\overset{\displaystyle H}{|}}{\underset{\underset{\displaystyle H}{|}}{C}}-\overset{\overset{\displaystyle H}{|}}{\underset{\underset{\displaystyle H}{|}}{C}}-O-H \longrightarrow H-\overset{\overset{\displaystyle H}{|}}{\underset{\underset{\displaystyle H}{|}}{C}}-\overset{\overset{\displaystyle O}{\|}}{C}-O-\overset{\overset{\displaystyle H}{|}}{\underset{\underset{\displaystyle H}{|}}{C}}-\overset{\overset{\displaystyle H}{|}}{\underset{\underset{\displaystyle H}{|}}{C}}-H \; + \; H_2O$$

(iv) State the reagents and conditions for the reaction of ethanol and ethanoic acid.

Explanation:

Concentrated H_2SO_4 as a catalyst with some heating are needed for the reaction between ethanol and ethanoic acid.

(v) Explain why the yield of the organic compound is not 100%.

Explanation:

The esterification process is a reversible reaction which does not go to completion. Hence, the yield of the organic compound is not 100%.

 Q Why does the concentrated H_2SO_4 catalyst not give us a 100% yield of the product?

A: The catalyst speeds up the rate and shortens the time to reach equilibrium, but it does not alter the yield of the product!

> (vi) Name the organic product and give one important industrial use for it.

Explanation:

The organic product is ethyl ethanoate and is a common industrial solvent.

Do you know?

— Some uses of ethanoic acid and esters:

- Ethanoic acid is used to make esters which can be a solvent for cosmetics, glues, etc. Sweet-smelling esters can also be artificial flavors for food.
- Ethanoic acid is used to manufacture insecticides and drugs.
- Hydrolysis of esters in animal fats or vegetable oils by sodium hydroxide is used to make soap and detergents.

---

7. (a) Compound **A** has the same empirical and molecular formulae. Elemental analysis of the compound shows it to have the following composition by mass: carbon, 38.10%; hydrogen, 7.41%; oxygen, 16.93%; and chlorine, 37.57%.

   (i) Explain the terms *empirical formula* and *molecular formula*.

Explanation:

The empirical formula (EF) of a compound is a representation of the *simplest* whole number ratio of the different atoms that made up the compound.

The molecular formula (MF) of a compound is a representation of the *actual* whole number ratio of the different atoms that made up the compound. Thus, MF = $n \times$ EF, i.e., an empirical formula is a fraction of the molecular formula.

(ii) Determine the molecular formula of compound **A**.

Explanation:

Elements present	C	H	O	Cl
Assuming 100 g of mass	38.10	7.41	16.93	37.57
Amount (in mol)	38.10/12 = 3.18	7.41/1 = 7.41	16.93/16 = 1.06	37.57/35.5 = 1.06
Divide by smallest number of mole	3.00	7.00	1.00	1.00
Simplest ratio	3	7	1	1

The empirical formula of the unknown compound is C_3H_7OCl. As the molecular formula is the same as the empirical formula, the molecular formula is C_3H_7OCl.

(iii) The compound contains a –OH group and a –Cl group. Draw two possible full constitutional/structural formulae for the compounds in which these groups are attached to different carbon atoms.

Explanation:

Q Are the three molecules constitutional/structural isomers?

A: Yes. Think further with these questions: (i) do they have the same chemical properties? If yes, why?; and (ii) do they have the same boiling points? If no, how would you separate them?

(iv) Suggest a simple chemical test for the presence of the –OH group.

Explanation:

Add a piece of sodium metal into compound **A**. The presence of the acidic –OH group would give an effervescence of H_2 gas. The H_2 gas would extinguish a lighted splint with a "pop" sound.

 Can we use hot $K_2Cr_2O_7$(aq) acidified with H_2SO_4(aq) to test for the presence of the –OH group?

A: Yes, you can do it here. The hot orange $K_2Cr_2O_7$(aq) that is acidified with H_2SO_4(aq) would be able to oxidize the –OH group with the $K_2Cr_2O_7$(aq) itself being reduced to the green Cr^{3+} ion.

 Is the $K_2Cr_2O_7$(aq) acidified with H_2SO_4(aq) used in the breathalyzer to test for ethanol?

A: Yes! Ethanol can be oxidized by the potassium dichromate(VI) and the amount of Cr^{3+} formed, which is indicated by the intensity of the green coloration, tells us the amount of ethanol in the exhaled air.

(b) Macromolecules are large molecules built up from many small units. Proteins and fats are natural macromolecules. Poly(chloroethene) and poly(ethene) are synthetic macromolecules.

(i) Name the type of linkage joining the units in fats. Suggest a method to form this linkage.

Explanation:

The type of linkage joining the units in fats is known as the ester linkage.

The linkage can be formed by reacting an alcohol with a carboxylic acid. The esterification process is a reversible reaction and it occurs at such a slow rate that equilibrium is attained only after a few hours. As such, concentrated H_2SO_4 or other acids, is used as a catalyst to speed up the reaction so as to shorten the time taken to reach equilibrium.

> (ii) Name the process for forming macromolecules of proteins and fats.

Explanation:

The process is known as condensation polymerization.

Do you know?

— One way of obtaining a polymer is through a condensation reaction in which monomers bind together with the *elimination* of small molecules, such as water or ammonia or hydrogen chloride. The polymer formed is called a *condensation polymer* and this process of forming the polymer is known as *condensation polymerization*.

— Here, we would just focus on two condensation polymers: the *polyester* and *polyamide* or *nylon*.

Formation of polyester
The following shows how polyester is formed generally from dicarboxylic acid and diol:

ester linkage

$$n \ HO-\overset{O}{\underset{\|}{C}}-R-\overset{O}{\underset{\|}{C}}-OH + n \ HO-R'-OH \longrightarrow \left[-\overset{O}{\underset{\|}{C}}-R-\overset{O}{\underset{\|}{C}}-O-R'-O- \right]_n + (2n-1) H_2O$$

dicarboxylic acid   diol   polyester

Poly(ethylene terephthalate) (commonly abbreviated as PET or PETE) is an example of a polyester that is made from benzene-1,4-dicarboxylic

(*Continued*)

(*Continued*)

acid and ethane-1,2-diol. Its name is derived from the old names of the respective monomers: terephthalic acid and ethylene glycol. Also known by its brand name, Terylene, this polyester is used in synthetic fibers. Its use in the textile industry accounts for the majority of the world's production of the polyester. Its other major use is in the plastic bottle production. The term "PET" is often used when talking about its use in packaging, whereas the term "polyester" is associative to its role as a synthetic fiber.

$$n \text{ HO—CH}_2\text{CH}_2\text{—OH} + n \text{ HO—C}{-}\!\!\bigcirc\!\!{-}\text{C—OH} \longrightarrow \left[\text{O—CH}_2\text{CH}_2\text{—O—C}{-}\!\!\bigcirc\!\!{-}\text{C}\right]_n + (2n-1)\,\text{H}_2\text{O}$$

| ethane-1,2-diol | benzene-1,4-dicarboxylic acid | | Terylene (one type of polyester) |

Formation of polyamide

Polyamides can be either natural or synthetic. The former includes proteins such as *silk* and *wool* while the latter includes *nylons*. Nylon is a generic term for a class of synthetic polyamides. Nylon is used in many applications. To name a few, nylon fibers are used to make fabrics, musical strings, and ropes. Solid nylon is used in automative and machine parts such as gears and bearings.

Belonging to this family is Nylon-6,6 whose name is based on the number of carbon atoms in the main chains of both the dicarboxylic acid and diamine monomers, i.e., hexanedioic acid and hexane-1,6-diamine.

$$n \text{ H}_2\text{N—(CH}_2)_6\text{—NH}_2 + n \text{ HO—C—(CH}_2)_4\text{—C—OH} \xrightarrow[\substack{270°\text{C} \\ 10\,\text{atm}}]{\text{heat}} \left[\text{N—(CH}_2)_6\text{—N—C—(CH}_2)_4\text{—C}\right]_n + (2n-1)\text{H}_2\text{O}$$

| hexane-1,6-diamine | hexanedioic acid | | nylon-6,6 |

(iii) Proteins can be hydrolyzed into monomers by boiling with concentrated hydrochloric acid. Name the monomers produced in this hydrolysis.

Explanation:

The monomers produced in this hydrolysis are known as amino acids.

> (iv) Hence, suggest why clothes made from nylon are damaged by concentrated hydrochloric acid.

Explanation:

Concentrated hydrochloric acid can catalyze the breakdown of the polymer into its monomers, hence damaging the structure of the clothes.

> (v) Poly(chloroethene) is made from the monomer chloroethene.
>
> (I) Name the process for forming poly(chloroethene).

Explanation:

The process is known as addition polymerization.

> (II) Draw the structure of poly(chloroethene) using three repeat units.

Explanation:

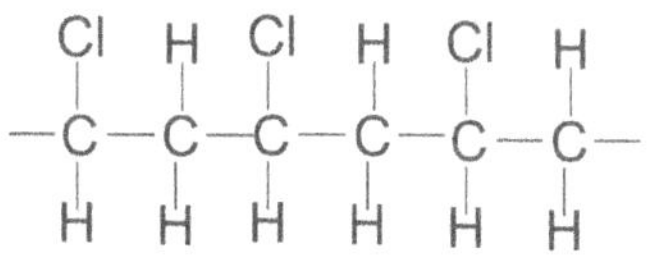

Q   Can we draw the following three repeat units instead?

$$
\begin{array}{ccccccc}
\text{H} & \text{Cl} & \text{H} & \text{Cl} & \text{H} & \text{Cl} \\
| & | & | & | & | & | \\
-\text{C}-\text{C}-\text{C}-\text{C}-\text{C}-\text{C}- \\
| & | & | & | & | & | \\
\text{H} & \text{H} & \text{H} & \text{H} & \text{H} & \text{H}
\end{array}
$$

A: Yes! If you want to know the difference between them, you can refer to *Understanding Advanced Organic and Analytical Chemistry* by K.S. Chan and J. Tan.

(III) Explain why poly(chloroethene) has a low melting point.

Explanation:

Poly(chloroethene) consists of polar macromolecules with pd–pd interactions between the molecules. These intermolecular forces are not very strong, hence accounting for the low melting point of the polymer.

(IV) Describe what you would observe when bromine reacts with chloroethene and state what type of reaction takes place.

Explanation:

As chloroethene contains a C=C double bond, it would be able to react with the chloroethene molecule through an addition reaction. Hence, the reddish-brown bromine would decolorize.

(V) Poly(chloroethene) is used as an insulating cover for electrical wires. Explain why poly(chloroethene) does not conduct electricity.

Explanation:

Poly(chloroethene) does not contain mobile charge carriers, hence it does not conduct electricity.

(VI) State and explain why plastics such as poly(ethene) may cause problems of environmental pollution.

Explanation:

Plastics such as poly(ethene) consist of long saturated chains which give the polymer a high degree of inertness since there are no reactive sites. This in turns accounts for the chemical resistance of all poly(alkenes) in general, when exposed to reagents such as acids and alkalis, which gives rise to their non-biodegradable nature.

(VII) What is the maximum mass of poly(ethene) that can be made from three tons of ethene?

Explanation:

Assuming none of the monomer is lost during the addition polymerization process, three tons of ethene would give three tons of poly(ethene).

Do you know?

— One disposal method for the non-biodegradable poly(alkenes) is to burn them, but chances are that it adds to air pollution. Incomplete combustion produces poisonous *carbon monoxide*. Those polymers that contain benzene (e.g., poly(styrene)) do not only produce large amounts of

(Continued)

(Continued)

soot, but also release poisonous *hydrocarbons*. When burned, PVC and other chlorinated polymers produce poisonous *hydrogen chloride* gas. These toxic fumes and many others that arise from the burning process are detrimental to both human health and the environment. The best, albeit dire, disposal method is basically dumping them into *landfills*, which poses a grim yet stark reality of our society's growing consumption of disposable plastics which are made to last. While scientists and companies are churning out biodegradable plastic products, the least we could do is to make a conscious effort in practicing the three Rs — reduce, reuse, and recycle.

CHAPTER 16

EXPERIMENTAL CHEMISTRY

1. (a) Suggest a separation method for the following:

   (i) Amino acids from a mixture obtained by the hydrolysis of a protein.

Explanation:

The amino acids can be separated by paper chromatography.

Do you know?

— The basic procedure of paper chromatography first involves using a pencil to draw a baseline on a rectangular strip of chromatography paper. The sample mixture of amino acids is spotted onto the baseline and the entire paper is then inserted upright into a beaker containing a suitable solvent, while making sure the baseline is above the solvent level. Immediately, the solvent creeps up the paper by *capillary action* and following its trail, are spots of amino acids moving at different speeds. Once the solvent is near the edge of the paper, the paper is removed and the solvent front is immediately marked out. The developed paper with the spots in their final positions is called the *chromatogram*. The separation

(Continued)

(Continued)

of the mixture of amino acids is very similar to the separation of ink in the following:

lid cover

chromatography
paper

glass
container

spot of black
ink (solute)

pencil
baseline

solvent

(a) at the start of the experiment

solvent front

coloured dyes start
to separate out

(b) during the experiment

solvent front

separated coloured
dyes (solutes)

(c) at the end of the separation

(ii) Naphthalene from a mixture of naphthalene and calcium fluoride.

Explanation:

As naphthalene sublimes readily, a mixture of naphthalene and calcium fluoride can be separated by heating the mixture in the following set-up.

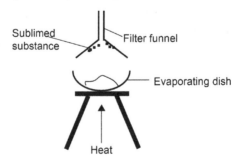

Sublimed
substance

Filter funnel

Evaporating dish

Heat

Solid naphthalene is collected on the inverted filter funnel, while calcium fluoride is left as residue in the evaporating dish.

(iii) Pure ethanol from a mixture of ethanol and solid camphor dissolved in it.

Explanation:

Ethanol can be separated from the mixture via simple distillation using the following set-up. The solid camphor would be left behind as residue while pure ethanol is collected as the distillate.

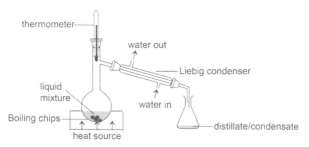

(iv) Oxygen from liquid air.

Explanation:

Oxygen can be separated from liquid air via fractional distillation.

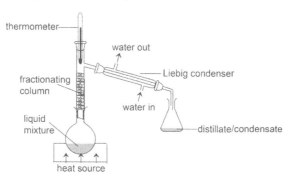

Do you know?

— Air is mainly made up of non-polar diatomic molecules such as oxygen and nitrogen. Although the intermolecular forces are weak, we can still liquefy air and *fractionally* separate it into various components. Air from the atmosphere needs to be filtered to remove the dust particles before being subjected to high pressure compression. The compression *releases* heat energy (remember when bonds form, energy is evolved?). This pressurized hot air is then suddenly allowed to expand into a very low pressure condition. The sudden expansion cools down the air. This compression and expansion processes are repeated until the air liquefies. The liquefied air is then fractionally distilled to separate it into liquid nitrogen, oxygen, and others.

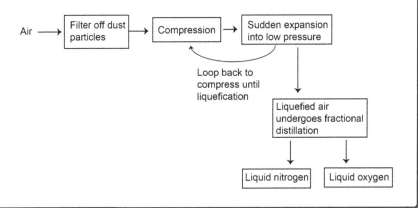

Q   Why does the sudden expansion cool down air?

A: During expansion, the distance of separation between the air particles increases. This means that energy is needed to *overcome* the intermolecular forces between the particles. Where does this needed energy come from? From the kinetic energy! Thus, with a lower kinetic energy, the temperature would be lower. Recall from Chapter 1 on the Particulate Nature of Matter, that average kinetic energy of a system of particles is directly proportional to the temperature.

(v) Calcium chloride from a mixture of calcium chloride and calcium carbonate.

Explanation:

The mixture of calcium chloride and calcium carbonate can be first dissolved in water. As calcium chloride is the only solid in the mixture that is soluble in water, the insoluble calcium carbonate can be separated via the following filtration:

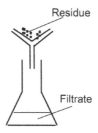

The calcium chloride can then be recovered by evaporating away the water as follows:

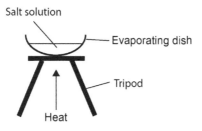

Q Can we use a beaker instead of an evaporating dish to evaporate the filtrate to dryness?

A: No. This is because it would take a longer time for the water to evaporate off due to the dripping back of the condensate on the side of the beaker.

An evaporating dish is more "open" and "shallower" than a beaker, and as such the water vapor would not be able to condense. This thus helps to save heat energy!

 Q   What should we do if the solid compound is prone to decomposition under strong heating? Can we still evaporate away the solvent using the above set-up?

A: Well, you can perform the evaporation on a steam bath as shown below:

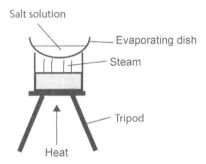

(vi) Barium sulfate from a mixture of barium sulfate and water.

Explanation:

Insoluble barium sulfate can be separated from a mixture of barium sulfate and water via filtration. The barium sulfate would be retained on the filter paper as residue, while the water is collected as the filtrate.

Q   How would you separate two immiscible liquids?

A: If we have two liquids, such as water and oil, that do not mix and are separated into two distinct layers, we say that they are *immiscible*. A separating funnel is used to separate such a liquid mixture.

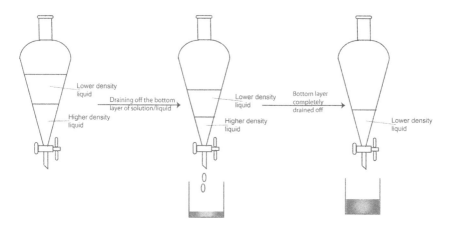

(b) For each of the following gas preparations:

- Give the names or formulae of suitable reactants.
- Give the drying agent needed.
- State how a gas jar full of the dry gas would be collected.

(i) Carbon dioxide;

Explanation:

CO_2 is produced by adding aqueous hydrochloric acid to calcium carbonate:

$$2HCl(aq) + CaCO_3(s) \rightarrow CaCl_2(aq) + CO_2(g) + H_2O(l).$$

Since CO_2 is acidic in nature, the gas can be dried by passing it through drying agents such as concentrated sulfuric(VI) acid, anhydrous calcium chloride, anhydrous copper(II) sulfate, and silica gel:

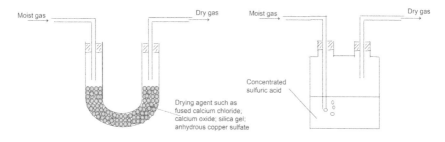

Since CO_2 is heavier than air, it can be collected by the upward displacement of air:

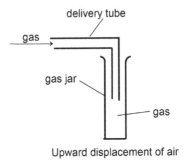

Upward displacement of air

Q   How would we know whether a gas is heavier than air?

A: Well, you can check its density. If not, the major components of air consist of nitrogen and oxygen molecules, which have molecular weights of 28 and 32, respectively. Now, CO_2 has molecular weight of 44. Thus, we expect CO_2 to be denser than air.

Q   Can we collect the CO_2 using a syringe?

A: Certainly you can. You can collect it using the following set-up:

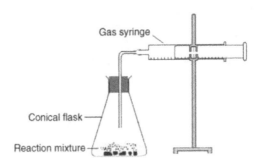

Q Can we collect CO_2 using the downward displacement of water?

A: This would not be too good as CO_2 is slightly soluble in water. Remember that CO_2 dissolves in water to give carbonic acid:

$$CO_2(g) + H_2O(l) \rightleftharpoons H_2CO_3(aq).$$

Q How can we test for the presence of CO_2 gas?

A: You can bubble the gas into limewater, $Ca(OH)_2(aq)$. If a white precipitate of $CaCO_3$ is formed, the gas is CO_2.

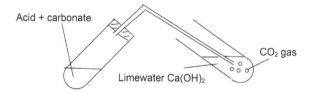

(ii) Ammonia;

Explanation:

Ammonia is produced by heating aqueous sodium hydroxide with ammonium nitrate:

$$NH_4NO_3(aq) + NaOH(aq) \rightarrow NH_3(aq) + H_2O(l) + NaNO_3(aq).$$

The ammonia gas produced is highly soluble in the aqueous medium. Hence, heating is needed to drive off the gas.

Since NH_3 is basic in nature, the gas can be dried by passing it through a basic drying agent such as calcium oxide.

Since NH_3 is lighter than air, it can be collected by the downward displacement of air:

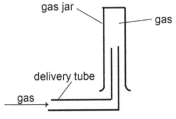

Downward displacement of air

Q Copper(II) sulfate is not an acidic compound. Why can't it be used to dry NH_3 gas?

A: You are right that copper(II) sulfate is not an acidic compound, but unfortunately, if you use copper(II) sulfate as the drying agent, the NH_3 gas would react with the Cu^{2+} ions as follows:

$$Cu^{2+} + 4NH_3 \rightarrow [Cu(NH_3)_4]^{2+},$$

to form a dark blue complex, $[Cu(NH_3)_4]^{2+}$.

Q Ammonia has a molecular weight of 17. So, does this cause it to be lighter than air?

A: Absolutely right!

Q How can we test for the presence of NH_3 gas?

A: Use a piece of moist red litmus paper. If it turns blue, the gas is ammonia as it is basic in nature.

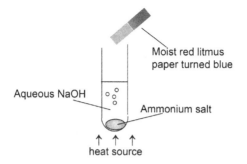

Q | Why must the litmus paper be moist?

A: An acid can only act as an acid ONLY when a base is present, and vice versa for a base. Water is acting as an acid, donating a proton to NH_3. It is the concentration of the free OH^- ions that causes the litmus paper to change colors:

$$NH_3(aq) \quad + \quad H_2O(l) \rightleftharpoons OH^-(aq) + NH_4^+(aq).$$

base	acid
proton acceptor	proton donor

(iii) Chlorine;

Explanation:

Chlorine is produced when concentrated sulfuric(VI) acid is added to sodium chloride in the presence of manganese(IV) oxide (MnO_2):

$$2NaCl(s) + 3H_2SO_4(aq) + MnO_2(s)$$
$$\rightarrow MnSO_4(aq) + 2NaHSO_4(aq) + 2H_2O(l) + Cl_2(g).$$

Since Cl_2 is acidic in nature, the gas can be dried by passing it through drying agents such as concentrated sulfuric(VI) acid, anhydrous calcium chloride, anhydrous copper(II) sulfate, and silica gel.

Since Cl_2 is heavier than air, it can be collected by the upward displacement of air.

Q   Can we collect the chlorine by the downward displacement of water?

A: The chlorine gas produced is relatively soluble in an aqueous medium. Hence, you can't collect it by the downward displacement of water. The solubility of chlorine in water is relatively high, i.e., the reason it is used to disinfect water.

Q   How can we test for the presence of chlorine gas?

A: Use a piece of moist litmus paper; it does not matter whether it is red or blue. The chlorine gas would bleach it. Or, you can use a piece of filter paper strip soaked with potassium iodide. If it turns from colorless to blue-black, the gas is Cl_2.

Q   Why did you use the "(aq)" symbol for concentrated H_2SO_4?

A: Well, there is still some water in the concentrated H_2SO_4 used.

(iv) Hydrogen;

Explanation:

Hydrogen is produced when zinc metal reacts with aqueous hydrochloric acid:

$$2HCl(aq) + Zn(s) \rightarrow ZnCl_2(aq) + H_2(g).$$

Since H_2 is neutral, the gas can be dried by passing it through drying agents such as concentrated sulfuric(VI) acid, anhydrous calcium chloride, anhydrous copper(II) sulfate, calcium oxide, and silica gel.

Since H_2 is lighter than air, it can be collected by the downward displacement of air.

Since H_2 is also insoluble in water, it can be collected by the downward displacement of water:

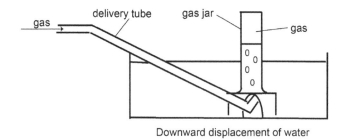

Downward displacement of water

Q Why is effervescence observed when hydrogen gas evolved?

A: Since the solubility of hydrogen in water is very low, effervescence would be observed.

Q How can we test for the presence of the hydrogen gas?

A: Use a lighted splint. If the flame is extinguished with a "pop" sound, the gas is hydrogen.

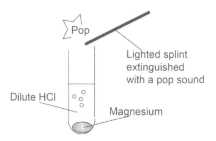

(v) Oxygen; and

Explanation:

Oxygen is produced when sodium nitrate or potassium nitrate is heated:

$$2NaNO_3(s) \rightarrow 2NaNO_2(s) + O_2(g).$$

Since O_2 is neutral, the gas can be dried by passing it through drying agents such as concentrated sulfuric(VI) acid, anhydrous calcium chloride, anhydrous copper(II) sulfate, calcium oxide, and silica gel.

Since O_2 is relatively insoluble in water, it can be collected by the downward displacement of water:

Q   How would you test for the presence of O_2 gas?

A: Use a glowing splint. If it is re-ignited, the gas is oxygen.

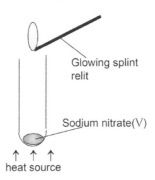

Glowing splint
relit

Sodium nitrate(V)

↑ ↑ ↑
heat source

(vi) Sulfur dioxide.

Explanation:

Sulfur dioxide is produced by adding aqueous hydrochloric acid to calcium sulfate(IV):

$$2HCl(aq) + CaSO_3(s) \rightarrow CaCl_2(aq) + SO_2(g) + H_2O(l).$$

Since SO_2 is acidic in nature, the gas can be dried by passing it through drying agents such as concentrated sulfuric(VI) acid, anhydrous calcium chloride, anhydrous copper(II) sulfate, and silica gel.

Since SO_2 is heavier than air, it can be collected by the upward displacement of air.

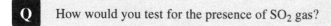

Q How would you test for the presence of SO_2 gas?

A: Use a piece of filter paper strip soaked with acidified potassium dichromate(VI). If it turns from orange to green, the gas is SO_2.

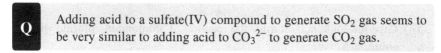

Q Adding acid to a sulfate(IV) compound to generate SO_2 gas seems to be very similar to adding acid to CO_3^{2-} to generate CO_2 gas.

A: Good observation! Both CO_3^{2-} and SO_3^{2-} are weak bases; hence, they can react with a strong acid.

Do you know?

— Summary for the test of gases:

Gas	Test and observation
Ammonia (NH_3)	Turns *moist* red litmus paper blue
Carbon dioxide (CO_2)	Forms a white precipitate with calcium hydroxide solution
Hydrogen (H_2)	Extinguishes lighted splint with a "pop" sound
Oxygen (O_2)	Relights a glowing splint
Sulfur dioxide (SO_2)	Turns aqueous potassium dichromate(VI) from orange to green
Chlorine (Cl_2)	Bleaches *moist* litmus paper

2. (a) When ammonium chloride reacts with solution **A** with warming, a colorless pungent gas **B** was evolved. When this gas is bubbled through dilute sulfuric acid, a colorless solution **C** is formed.

   (i) State the identity of solutions **A** and **C**.

Explanations:

Solutions **A** and **C** are aqueous NaOH and aqueous ammonium sulfate, respectively.

(ii) Give the ionic equation for the reaction of ammonium chloride with solution **A**.

Explanations:

$$NH_4^+(aq) + OH^-(aq) \rightarrow NH_3(g) + H_2O(l).$$

(iii) Explain why warming is necessary.

Explanation:

The ammonia gas produced is highly soluble in the aqueous medium. Hence, heating is needed to drive off the gas.

(iv) Give a balanced equation for the reaction between gas **B** and dilute sulfuric acid.

Explanation:

$$2NH_3(g) + H_2SO_4(aq) \rightarrow (NH_4)_2SO_4(aq).$$

(v) State a test for gas **B**. Give the expected observation.

Explanation:

Use a piece of moist red litmus paper. If it turns blue, the gas is ammonia as it is basic in nature.

(vi) If gas **B** is dry, it has no effect on the test mentioned in part (a)(v). Explain why it is so.

Explanation:

If the NH_3 is dry, then the NH_3 cannot hydrolyze in water as follows:

$$NH_3(aq) + H_2O(l) \rightleftharpoons OH^-(aq) + NH_4^+(aq).$$

Without the OH^- ions being generated, the pH of the solution would not change. Hence, the red litmus paper would not change color.

(b) When copper(II) carbonate is heated, a colorless gas **D** is evolved. When dilute hydrochloric acid is added to the residue after heating, a blue solution, **E**, is obtained. When dilute sulfuric acid is added to solid copper(II) carbonate, effervescence of gas **D** is observed and a solution **F** is obtained. Addition of aqueous ammonia to **F** gives a blue precipitate which dissolves in excess aqueous ammonia, giving a dark-blue solution.

(i) Name gas **D** and give a chemical test to identify the gas.

Explanation:

Gas **D** is carbon dioxide.

Bubble the gas into limewater, $Ca(OH)_2(aq)$; if a white precipitate of $CaCO_3$ is formed, the gas is CO_2.

> **Q**   What type of reactions would cause the formation of a precipitate?

A: Basically, a precipitate may form when two clear solutions (i.e., without any solid present) are mixed. The formation of a precipitate indicates that the solid that is formed must be insoluble in water. So first, you need to know the insoluble compounds. The following table can help you do that:

1   All carbonates are insoluble in water, except for sodium carbonate, potassium carbonate, and ammonium carbonate.

2   All hydroxides are insoluble in water, except for sodium hydroxide, potassium hydroxide and barium hydroxide. Calcium hydroxide is sparingly soluble.

3   Barium sulfate, lead(II) sulfate, and calcium sulfate are insoluble in water.

4   Lead(II) chloride and silver chloride are insoluble in water. Similarly, lead(II) bromide and silver bromide are also insoluble in water.

5   Lead(II) iodide and silver iodide are insoluble in water.

> (ii)   Give a balanced equation for the heating of solid copper(II) carbonate with state symbols.

Explanation:

$$CuCO_3(s) \rightarrow CuO(s) + CO_2(g).$$

> **Q**   So, the heating of a solid can actually help us identify the type of compound that is present?

A: Yes! The heating of a solid to test for gas is part of the qualitative analysis of a compound. Qualitative analysis is a branch of analytical chemistry which only involves knowing the *identity* of the compound. If you have a mixture containing some anions or cations or both, how are you going to systematically determine each of their identities? Is there any specific test that would help you to pinpoint the identities of the species that are present?

 Q So, what is considered a good test in qualitative analysis?

A: In qualitative analysis, a good test must be one that gives us a visual observation: color change, evolution of a gas, or formation of a precipitate. For example, the following table lists the characteristic tests for the presence of anions and cations:

Anion	Characteristic test	Observation
Carbonate (CO_3^{2-})	Add dilute acid.	Effervescence of CO_2 gas evolved.
Chloride (Cl^-)	Acidify with dilute nitric acid, follow by aqueous silver nitrate.	White ppt (AgCl).
Iodide (I^-)	Acidify with dilute nitric acid, follow by aqueous lead(II) nitrate.	Yellow ppt (PbI_2).
Nitrate (NO_3^-)	Add aqueous sodium hydroxide follow by Devarda's alloy (aluminum foil) and then warm gently.	Ammonia gas evolved
Sulfate (SO_4^{2-})	Acidify with dilute nitric acid, follow by aqueous barium nitrate.	White ppt ($BaSO_4$).

Q Why must we acidify the sample with dilute nitric acid first before adding silver nitrate, lead(II) nitrate, or barium nitrate?

A: If there is any carbonate ion present in the sample, it would be decomposed by the addition of dilute nitric acid. Hence, any precipitate that is subsequently formed would be due to the anion that is to be tested and not due to the carbonate ion.

Q Can we use $H_2SO_4(aq)$ or HCl(aq) to do the acidification?

A: No! This because $H_2SO_4(aq)$ or HCl(aq) would introduce SO_4^{2-} and Cl^- ions, respectively. Thus, if there is a precipitate formed, it might be due to a cation that is present in the sample which is able to form a precipitate with the SO_4^{2-} or Cl^- ion that is being introduced. So, using nitric acid is a good choice because "all nitrates are soluble".

 Q So, barium nitrate instead of barium chloride is used to test for SO_4^{2-} ion because we need to prevent the introduction of the Cl^- ion which may form a precipitate with a cation in the sample?

A: Absolutely right!

Cation	Adding aqueous sodium hydroxide	Adding aqueous ammonia
Aluminum (Al^{3+})	White ppt soluble in excess to give a colorless solution. $Al^{3+}(aq) + 3OH^-(aq) \rightarrow Al(OH)_3(s)$ $Al(OH)_3(s) + OH^-(aq) \rightarrow$ $[Al(OH)_4]^-(aq)$	White ppt insoluble in excess. $Al^{3+}(aq) + 3OH^-(aq) \rightarrow Al(OH)_3(s)$
Ammonium (NH_4^+)	Ammonia evolved on warming $NH_4^+(aq) + OH^-(aq) \rightarrow NH_3(g) +$ $H_2O(l)$	—
Calcium (Ca^{2+})	White ppt insoluble in excess. $Ca^{2+}(aq) + 2OH^-(aq) \rightarrow Ca(OH)_2(s)$	No ppt.
Copper(II) (Cu^{2+})	Blue ppt insoluble in excess. $Cu^{2+}(aq) + 2OH^-(aq) \rightarrow Cu(OH)_2(s)$	Blue ppt soluble in excess to give a dark blue solution. $Cu^{2+}(aq) + 2OH^-(aq) \rightarrow$ $Cu(OH)_2(s)$ $Cu(OH)_2(s) + 4NH_3(aq) \rightarrow$ $[Cu(NH_3)_4]^{2+}(aq) + 2OH^-(aq)$
Iron(II) (Fe^{2+})	Green ppt insoluble in excess. $Fe^{2+}(aq) + 2OH^-(aq) \rightarrow Fe(OH)_2(s)$	Green ppt insoluble in excess. $Fe^{2+}(aq) + 2OH^-(aq) \rightarrow Fe(OH)_2(s)$
Iron(III) (Fe^{3+})	Reddish-brown ppt insoluble in excess. $Fe^{3+}(aq) + 3OH^-(aq) \rightarrow Fe(OH)_3(s)$	Reddish-brown ppt insoluble in excess. $Fe^{3+}(aq) + 3OH^-(aq) \rightarrow Fe(OH)_3(s)$
Lead(II) (Pb^{2+})	White ppt soluble in excess to give a colorless solution. $Pb^{2+}(aq) + 2OH^-(aq) \rightarrow Pb(OH)_2(s)$ $Pb(OH)_2(s) + 2OH^-(aq) \rightarrow$ $[Pb(OH)_4]^{2-}(aq)$	White ppt insoluble in excess. $Pb^{2+}(aq) + 2OH^-(aq) \rightarrow Pb(OH)_2(s)$
Zinc (Zn^{2+})	White ppt soluble in excess to give a colorless solution. $Zn^{2+}(aq) + 2OH^-(aq) \rightarrow Zn(OH)_2(s)$ $Zn(OH)_2(s) + 2OH^-(aq) \rightarrow$ $[Zn(OH)_4]^{2-}(aq)$	White ppt soluble in excess to give a colorless solution. $Zn^{2+}(aq) + 2OH^-(aq) \rightarrow Zn(OH)_2(s)$ $Zn(OH)_2(s) + 4NH_3(aq) \rightarrow$ $[Zn(NH_3)_4]^{2+}(aq) + 2OH^-(aq)$

Q Why does Ca^{2+} form a precipitate with aqueous NaOH but not $NH_3(aq)$?

A: This is because $NH_3(aq)$ is a weaker base compared to aqueous NaOH. Hence, the concentration of OH^- in $NH_3(aq)$ is too low to cause the $Ca(OH)_2$ to be precipitated out. For further details, you can refer to *Understanding Advanced Physical Inorganic Chemistry* by J. Tan and K.S. Chan.

(iii) Give an ionic equation for the reaction of dilute hydrochloric acid with the residue.

Explanation:

$$CuO(s) + 2H^+(aq) \rightarrow Cu^{2+}(aq) + H_2O(l).$$

(iv) Explain why effervescence was observed.

Explanation:

Not all the solid carbonate have been decomposed, the residual carbonate reacted with the acid. Since the solubility of carbon dioxide in water is relatively low, effervescence was observed.

(v) Give an ionic equation for the formation of the blue precipitate.

Explanation:

$$Cu^{2+}(aq) + 2OH^-(aq) \rightarrow Cu(OH)_2(s).$$

(vi) Give a balanced equation for the dissolution of the precipitate to form a dark blue solution.

Explanation:

$$Cu(OH)_2(s) + 4NH_3(aq) \rightarrow [Cu(NH_3)_4]^{2+}(aq) + 2OH^-(aq).$$

3. State whether each of the following mixtures will undergo a chemical change. If it does, write a balanced equation.

Mixture	Reaction?	Equation
$NaOH(aq) + ZnSO_4(aq)$		
$HCl(aq) + H_2SO_4(aq)$		
$AgNO_3(aq) + KCl(aq)$		
$Na_2CO_3(aq) + CuCl_2(aq)$		
$NaCl(aq) + KOH(aq)$		
$Ca(NO_3)_2(aq) + FeCl_2(aq)$		
$Ca(NO_3)_2(aq) + Na_2CO_3(aq)$		
$HCl(aq) + BaCO_3(s)$		
$H_2SO_4(aq) + Ba(NO_3)_2(aq)$		
$H_2SO_4(aq) + Zn(s)$		

Explanation:

Mixture	Reaction?	Equation
$NaOH(aq) + ZnSO_4(aq)$	Yes	White ppt soluble in excess $NaOH(aq)$ to give a colorless solution. $ZnSO_4(aq) + 2NaOH(aq) \rightarrow Zn(OH)_2(s) + Na_2SO_4(aq)$ $Zn(OH)_2(s) + 2NaOH(aq) \rightarrow Na_2[Zn(OH)_4](aq)$
$HCl(aq) + H_2SO_4(aq)$	No	—
$AgNO_3(aq) + KCl(aq)$	Yes	White ppt (AgCl) is formed. $AgNO_3(aq) + KCl(aq) \rightarrow AgCl(s) + KNO_3(aq)$
$Na_2CO_3(aq) + CuCl_2(aq)$	Yes	Green ppt ($CuCO_3$) is formed. $Na_2CO_3(aq) + CuCl_2(aq) \rightarrow CuCO_3(s) + 2NaCl(aq)$
$NaCl(aq) + KOH(aq)$	No	—
$Ca(NO_3)_2(aq) + FeCl_2(aq)$	No	—
$Ca(NO_3)_2(aq) + Na_2CO_3(aq)$	Yes	White ppt ($CaCO_3$) is formed. $Ca(NO_3)_2(aq) + Na_2CO_3(aq) \rightarrow CaCO_3(s) + 2NaNO_3(aq)$
$HCl(aq) + BaCO_3(s)$	Yes	Effervescence of CO_2 gas is observed. $2HCl(aq) + BaCO_3(s) \rightarrow BaCl_2(aq) + CO_2(g) + H_2O(l)$
$H_2SO_4(aq) + Ba(NO_3)_2(aq)$	Yes	White ppt ($BaSO_4$) is formed. $H_2SO_4(aq) + Ba(NO_3)_2(aq) \rightarrow BaSO_4(s) + 2HNO_3(aq)$
$H_2SO_4(aq) + Zn(s)$	Yes	Effervescence of H_2 gas is observed. $H_2SO_4(aq) + Zn(s) \rightarrow ZnSO_4(aq) + H_2(g)$

4. You are provided with a solution **FA 1** containing the cations $Al^{3+}(aq)$, $Fe^{3+}(aq)$, and $Cu^{2+}(aq)$.

   (a) (i) What is the color of solution **FA 1**?

Explanation:

The colors of the cations $Al^{3+}(aq)$, $Fe^{3+}(aq)$, and $Cu^{2+}(aq)$ are colorless, yellow, and blue, respectively. Hence, a solution of **FA 1** would look green (yellow + blue) in color.

 Q So, the color of the unknown compound could help us in the identification process?

A: Certainly! Certain compounds are colored because of the presence of a specific cation or anion. So, it is good to know some of their colors. The following tables would be helpful:

Color of solids/ precipitates	Possible identity
Black	CuO, MnO_2, I_2, and metal powder like Fe filings.
Blue	Copper(II) compounds, e.g., $CuSO_4$, $Cu(NO_3)_2$, and $Cu(OH)_2$.
Reddish-brown	$Fe(OH)_3$ and copper metal.
Green	$CuCO_3$ and $CuCl_2$. $FeSO_4$ is pale green; $Fe(OH)_2$ is green.
Orange	Potassium dichromate(VI), $K_2Cr_2O_7$.
Yellow	PbI_2 and AgI are pale yellow.
Purple	Potassium manganate(VII), $KMnO_4$.
White	Most compounds of Na^+, K^+, Ca^{2+}, Zn^{2+}, Al^{3+}, Pb^{2+}, $Mg2^+$ and NH_4^+. Important ones include $BaSO_4$, $PbSO_4$, $CaSO_4$, $PbCl_2$, and AgCl. {ZnO is yellow when hot and white when cooled.}

Color of liquids/ solutions	Possible identity
Blue	Aqueous copper(II) compounds.
Brown	Aqueous I_2 with aqueous KI.
Orange	Aqueous $K_2Cr_2O_7$. Bromine dissolves in organic solvents (e.g., hexane).

(Continued)

(*Continued*)

Color of liquids/ solutions	Possible identity
Yellow	Aqueous $FeCl_3$ (dark yellow).
	Aqueous Br_2.
	Aqueous I_2 (very pale yellow).
Purple	Aqueous $KMnO_4$.
Violet	Iodine dissolves in organic solvents (e.g., hexane) to form a violet solution.

 Q   I assumed solid iodine is not soluble in water, so why is aqueous iodine with KI brown in color?

A: Iodine indeed is not very soluble in water. A solution containing iodine is extremely pale yellow in color. But what happens when KI is added? The following reaction takes place:

$$I_2(aq) + I^-(l) \rightleftharpoons I_3^-(aq).$$

As $I_3^-(aq)$ is an ion, it can interact well with the water molecules. This hence increases the solubility of iodine in the aqueous KI solution.

(ii) How does the color of solution **FA 1** indicate that $Fe^{3+}(aq)$ and $Cu^{2+}(aq)$ cations quoted are both present?

Explanation:

The $Fe^{2+}(aq)$ ion in water is pale green in color. The green coloration, like that of green grass, observed for **FA 1** is a combination of yellow $Fe^{3+}(aq)$ and blue $Cu^{2+}(aq)$ and not because of the $Fe^{2+}(aq)$ ion.

(b) State with reasons, whether each of the following anions could also be present.
  (i) $NO_3^-(aq)$; and

Explanation:

The NO_3^-(aq) ion can be present as it does not form any insoluble compounds with the cations that are present in **FA 1**.

(ii) SO_4^{2-}(aq).

Explanation:

The SO_4^{2-}(aq) ion can be present as it does not form any insoluble compounds with the cations that are present in **FA 1**.

Q   So, can the anions CO_3^{2-}, Cl^-, and I^- be also present in **FA 1**?

A: The Cl^- anion can be present. But both the CO_3^{2-} and I^- anions cannot be present in **FA 1** due to the following reasons:

— As **FA 1** is a solution, the CO_3^{2-} anion cannot be present as it would form an insoluble $CuCO_3$ solid with the Cu^{2+} ion. In addition, a solution containing Al^{3+}(aq) and Fe^{3+}(aq) is acidic due to the high charge densities of the triply charged cations. These highly charged cations would polarize the electron cloud of the H_2O to an extent that the O–H bonds are weakened and can be broken easily. The H^+ ions that are formed would decompose the CO_3^{2-} anion. This would mean that $Fe_2(CO_3)_3$ and $Al_2(CO_3)_3$ do not exist at all.

— In the presence of I^-, Fe^{3+} would undergo the following redox reaction:

$$2Fe^{3+}(aq) + 2I^-(aq) \rightarrow 2Fe^{2+}(aq) + I_2(aq).$$

Thus, FeI_3 does not exist at all!

(c) Describe briefly the test that could be carried out to determine the presence of SO_4^{2-}(aq).

Explanation:

Add aqueous barium nitrate to the unknown solution. If a white precipitate of $BaSO_4$ is formed, the SO_4^{2-}(aq) ion is present.

> **Q**   Can we use aqueous barium chloride?

A: Yes, you can, provided your unknown solution does not contain the Ag^+ ion as insoluble AgCl can be precipitated out by the Cl^- ion from the barium chloride.

> **Q**   Can we use aqueous lead nitrate?

A: Yes, you can, provided your unknown solution does not contain the Cl^- ion as insoluble $PbCl_2$ can be precipitated out by the Pb^{2+} ion from the lead nitrate.

> **Q**   Is there a way to systematically identify the cation or the anion that is present in the unknown solution?

A: Yes! To identify the unknown anion, you can follow the following sequence of tests:

Step 1:   Add dilute H_2SO_4(aq)/HNO_3(aq)/HCl(aq) to a fresh sample.

Fresh sample ⟶ Add dilute H_2SO_4

- Gas evolved — CO_3^{2-} present
- No gas — I^-, Cl^-, NO_3^-, SO_4^{2-} present

Step 2:   Add $BaCl_2$(aq)/$Ba(NO_3)_2$(aq) to a fresh sample.

Fresh sample ⟶ Add dilute $BaCl_2$/ $Ba(NO_3)_2$

- White ppt — SO_4^{2-} present
- No white ppt — I^-, Cl^-, NO_3^- present

Step 3:   Add $Pb(NO_3)_2$(aq) to a fresh sample.

Fresh sample ⟶ Add dilute $Pb(NO_3)_2$

- Have ppt — White ppt ($PbCl_2$) / Yellow ppt (PbI_2)
- No ppt — NO_3^- present

Step 4:   Add NaOH(aq) to a fresh sample and warm, followed by addition of Devarda's alloy.

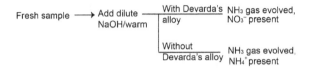

To identify the cation, you can use the following sequence of steps:

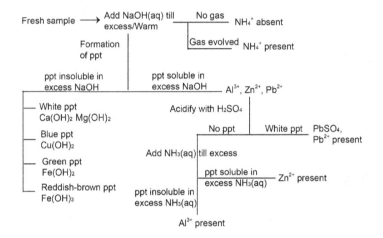

But if you prefer to start with NH_3(aq) as the testing reagent, then you may carry out the following procedures in accordance to the flowchart:

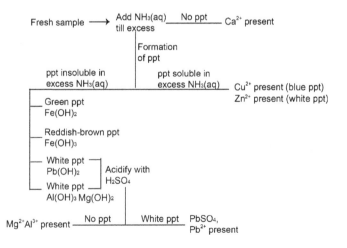

(d) You are provided with the reagents NaOH(aq), NH$_3$(aq), HCl(aq), and distilled water together with test-tubes, a filter funnel, and filter papers. Devise a sequence of steps by which, using **only** these materials, the three cations could be separately precipitated, i.e., so that each precipitate contains only one cation. Give a description of your proposed sequence of steps below, numbering each step.

Explanation:

(i) Add NaOH(aq) to the mixture till it is in excess.

— $Fe^{3+}(aq) + 3OH^-(aq) \rightarrow Fe(OH)_3(s)$
$Cu^{2+}(aq) + 2OH^-(aq) \rightarrow Cu(OH)_2(s)$
$Al^{3+}(aq) + 3OH^-(aq) \rightarrow Al(OH)_3(s) + OH^-(aq) \rightarrow$
$[Al(OH)_4]^-(aq).$

(ii) Filter the mixture.

— The residue would contain the insoluble $Fe(OH)_3$ and $Cu(OH)_2$ while the filtrate would contain the $[Al(OH)_4]^-$ complex.

(iii) Add HCl(aq) separately to both the residue and the filtrate in part (ii).

— Residue: $Fe(OH)_3(s) + 3H^+(aq) \rightarrow Fe^{3+}(aq) + 3H_2O(l)$
$Cu(OH)_2(s) + 2H^+(aq) \rightarrow Cu^{2+}(aq) + 2H_2O(l).$
— Filtrate: $[Al(OH)_4]^-(aq) + 4H^+(aq) \rightarrow Al^{3+}(aq) + 4H_2O(l).$

(iv) Add NH$_3$(aq) separately to both the residue and the filtrate in part (iii) till excess NH$_3$(aq) is added.

— Residue: $Fe^{3+}(aq) + 3OH^-(aq) \rightarrow Fe(OH)_3(s)$
$Cu^{2+}(aq) + 2OH^-(aq) \rightarrow Cu(OH)_2(s)$
$Cu(OH)_2(s) + 4NH_3(aq) \rightarrow [Cu(NH_3)_4]^{2+}(aq) +$
$2OH^-(aq).$
— Filtrate: $Al^{3+}(aq) + 3OH^-(aq) \rightarrow Al(OH)_3(s).$

(v) Filter the residue mixture in part (iv).

— The residue would contain the insoluble $Fe(OH)_3$ while the filtrate would contain the $[Cu(NH_3)_4]^{2+}$ complex.

(vi) Add NaOH(aq) to the filtrate in part (v) till excess NaOH(aq) is added.

— $[Cu(NH_3)_4]^{2+}(aq) + 2OH^-(aq) \rightarrow Cu(OH)_2(s) + 4NH_3(aq).$

5. The following is a typical procedure to identify the cations and anions that are present in a solution. Fill in the expected observation and deduction in the grey boxes and deduce the cation (two only without Mg^{2+}) and anion (two only) that are present in the solution.

Test	Observation	Deduction
(a) Add dilute H_2SO_4(aq) to **FA 1**.		Ba^{2+} and Pb^{2+} are absent. CO_3^{2-}, SO_3^{2-}, and NO_2^- are absent. Colored ions are absent.
(b) Add NaOH(aq) to **FA 1**. Filter and retain filtrate for test **(c)** and **(d)**.	White ppt formed, insoluble in excess NaOH(aq). Another white ppt observed, soluble in excess NaOH(aq) to give a colorless solution. Colorless filtrate.	
(c) Add HNO_3(aq) to a portion of filtrate from test (b). Divide into 3 portions.	White ppt formed, dissolved upon further addition of HNO_3(aq).	$Al(OH)_3$(s)/ $Zn(OH)_2$(s) Al^{3+}(aq) or Zn^{2+}(aq) is present.
(i) To 1st portion, add $Ba(NO_3)_2$(aq).		SO_4^{2-}(aq) is absent.
(ii) To 2nd portion, add $AgNO_3$(aq).	White ppt formed.	
(iii) To 3rd portion, add NH_3(aq).	White ppt formed, soluble in excess NH_3(aq), giving a colorless solution.	
(d) Warm a portion of the filtrate from **(b)**. Cool, then add Al powder and warm cautiously.	No gas evolved that turned moist red litmus blue. Pungent gas evolved turned moist red litmus blue.	
(e) Add NH_3(aq) to **FA 1**.	White ppt formed, soluble in excess NH_3(aq).	

Explanation:

Test	Observation	Deduction
(a) Add dilute $H_2SO_4(aq)$ to **FA 1**.	No ppt formed. No gas evolved. FA 1 is not a colored solution.	Ba^{2+} and Pb^{2+} are absent. CO_3^{2-}, SO_3^{2-}, and NO_2^- are absent. Colored ions are absent.
(b) Add NaOH(aq) to **FA 1**. Filter and retain filtrate for test **(c)** and **(d)**.	White ppt formed, insoluble in excess NaOH(aq). Another white ppt observed, soluble in excess NaOH(aq) to give a colorless solution. Colorless filtrate.	$Ca(OH)_2(s)$ $Ca^{2+}(aq)$ is present. $Al(OH)_3(s)/Zn(OH)_2(s)$ $Al^{3+}(aq)$ or $Zn^{2+}(aq)$ are present. $[Al(OH)_4]^-(aq)/$ $[Zn(OH)_4]^{2-}(aq)$
(c) Add $HNO_3(aq)$ to a portion of filtrate from test **(b)**. Divide into 3 portions.	White ppt formed, dissolved upon further addition of $HNO_3(aq)$.	$Al(OH)_3(s)/Zn(OH)_2(s)$ $Al^{3+}(aq)$ or $Zn^{2+}(aq)$ are present.
(i) To 1st portion, add $Ba(NO_3)_2(aq)$.	No white ppt formed.	$SO_4^{2-}(aq)$ is absent.
(ii) To 2nd portion, add $AgNO_3(aq)$.	White ppt formed.	$AgCl(s)$. $Cl^-(aq)$ is present.
(iii) To 3rd portion, add $NH_3(aq)$.	White ppt formed, soluble in excess $NH_3(aq)$, giving a colorless solution.	$Zn(OH)_2(s)$. $Zn^{2+}(aq)$ is present.
(d) Warm a portion of the filtrate from **(b)**. Cool, then add Al powder and warm cautiously.	No gas that turned moist red litmus blue evolved. Pungent gas evolved turned moist red litmus blue.	$NH_4^+(aq)$ is absent. $NH_3(g)$ evolved. $NO_3^-(aq)$ is present.
(e) Add $NH_3(aq)$ to **FA 1**.	White ppt formed, soluble in excess $NH_3(aq)$.	$Zn(OH)_2(s)$ $Zn^{2+}(aq)$ is present.

Cations present: Ca^{2+} and Zn^{2+}.
Anions present: Cl^- and NO_3^-.

Q   If you suspect that a gas is evolved from a reaction, how can you sequentially identify the gas?

A: Well, you can follow the following guidelines:

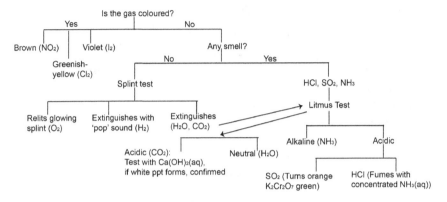

INDEX